FISH, FISHING AND FISHERIES

INTERNATIONAL FISHERIES MANAGEMENT

IMPROVEMENT EFFORTS AS REPORTED TO CONGRESS

FISH, FISHING AND FISHERIES

Additional books in this series can be found on Nova's website under the Series tab.

Additional e-books in this series can be found on Nova's website under the e-book tab.

FISH, FISHING AND FISHERIES

INTERNATIONAL FISHERIES MANAGEMENT

IMPROVEMENT EFFORTS AS REPORTED TO CONGRESS

MARLIN D. ANDERSON
EDITOR

nova publishers
New York

Library of Congress Cataloging-in-Publication Data

ISBN: 978-1-63483-170-3

Published by Nova Science Publishers, Inc. † New York

CONTENTS

PREFACE

In 2006 Congress recognized the need for international cooperation to address some of the most significant issues affecting international fisheries today: illegal, unreported, and unregulated (IUU) fishing and fishing practices that may undermine the sustainability of living marine resources. In 2010, Congress focused on the need for enhanced international action to conserve and protect sharks in passing the Shark Conservation Act. By requirement, the Secretary of Commerce, in biennial reports, is to identify nations whose fishing vessels are engaged in certain IUU fishing, bycatch, and shark fishing practices; describe U.S. consultations with the identified countries to urge appropriate actions; and certify whether such actions subsequent to identification have adequately addressed the offending activities. The Secretary of Commerce has delegated authority to identify and certify countries under the Moratorium Protection Act to the NOAA Assistant Administrator for Fisheries. The Secretary of Commerce is also directed, in consultation with the Secretary of State and in cooperation with relevant regional fishery management councils and any relevant advisory committees, to take certain actions to improve the effectiveness of international fishery management organizations in conserving and managing stocks under their jurisdiction. This book surveys efforts by the United States to strengthen its leadership toward improving international fisheries management and enforcement, particularly with regard to IUU fishing, bycatch of protected living marine resources (PLMRs), and certain shark fishing practices. Furthermore, this book covers issues that have been identified during congressional hearings and in legislation introduced during the last three Congresses, some of which include the flexibility in rebuilding overfished fisheries; annual catch limits; uncertainty and data needs; catch shares (limited access privilege programs); management process and decision making; bycatch; and environmental quality.

In: International Fisheries Management
Editor: Marlin D. Anderson
ISBN: 978-1-63483-170-3

Chapter 1

IMPROVING INTERNATIONAL FISHERIES MANAGEMENT*

U.S. Department of Commerce

EXECUTIVE SUMMARY

The Magnuson-Stevens Fishery Conservation and Management Reauthorization Act of 2006 (MSRA) acknowledged the need for international cooperation to address fishing activities that have a deleterious effect on sustainable fisheries worldwide. In amending the High Seas Driftnet Fishing Moratorium Protection Act (Moratorium Protection Act), Congress directed the Executive Branch to strengthen its leadership in international fisheries management and enforcement, particularly with regard to illegal, unreported, and unregulated (IUU) fishing and unsustainable fishing practices such as bycatch of protected living marine resources (PLMRs). The Shark Conservation Act of 2010 (SCA) amended the Moratorium Protection Act to add a third focus: directed and incidental catch of sharks, especially the practice of finning, in areas beyond national jurisdiction. Congress requires the Secretary of Commerce to identify countries whose fishing vessels were engaged in these activities, and to consult with those countries on improving their fisheries management and enforcement practices.

* This is an edited, reformatted and augmented version of a report to Congress, issued by the National Marine Fisheries Service, February 2015.

In its 2013 Report to Congress, the National Marine Fisheries Service (NMFS), a line office of the National Oceanic and Atmospheric Administration (NOAA) in the Department of Commerce, identified 10 countries as having vessels engaged in IUU fishing during the preceding two years: Colombia, Ecuador, Ghana, Italy, Mexico, Panama, the Republic of Korea, Spain, Tanzania, and Venezuela. Mexico was also identified for bycatch of North Pacific loggerhead sea turtles, the first time a country has been identified under the bycatch provision of the Moratorium Protection Act. This report details the consultations with those countries over the past two years. As required by the MSRA, it contains NMFS' certification decisions for those 10 countries, in Part IV. NMFS is making positive certifications for all of the countries identified for IUU fishing; however, NMFS is delaying Mexico's certification determination for bycatch until May 2015.

In this report, NMFS identifies four countries as having been engaged in IUU fishing based on violations of international conservation and management measures (CMMs) during 2013 and/or 2014: Colombia, Ecuador, Nicaragua, and Portugal. NMFS also identifies Mexico as having been engaged in IUU fishing based on overfishing shared stocks and fishing in the U.S. Exclusive Economic Zone (EEZ). Nigeria is identified as having been engaged in IUU fishing that undermined the conservation measures of a regional fisheries management organization (RFMO) of which it is not a member. NMFS considered five other countries for identification for IUU fishing during the reporting period, but consultations indicate corrective actions have already been taken to address the fishing activities of concern, or the allegations of IUU fishing activities were refuted. No countries are identified for bycatch of PLMRs or for shark catch on the high seas, due primarily to the restrictive time frames and other limitations in the statute.

In addition, this report updates domestic, regional, and global efforts to combat IUU fishing, minimize bycatch of protected species, and conserve sharks. Among the most important developments during the past two years are the following:

- On April 3, 2014, the Senate gave its advice and consent to ratification of the 2009 Agreement on Port State Measures, which requires parties to take actions to prevent IUU fish and fish products from entering the stream of commerce. U.S. ratification of this agreement is pending enactment of implementing legislation.
- On that same day, the Senate gave its advice and consent to ratification of three other fisheries-related treaties: two establishing

the South Pacific Regional Fisheries Management Organization and the North Pacific Fisheries Commission, and one amending the treaty for the management of fisheries in the North Atlantic. U.S. ratification of these agreements is pending enactment of implementing legislation.

- Several species of sharks threatened by international trade gained new global protection. The 16th Conference of the Parties to the Convention on International Trade in Endangered Species of Wild Fauna and Flora (CITES) agreed to list the oceanic whitetip shark, three species of hammerhead sharks (scalloped, smooth, and great), and the porbeagle shark, as well as manta rays, in CITES Appendix II – a step that means increased protection, but still allows for legal and sustainable trade.
- In June 2014, President Obama directed establishment of a task force on combating IUU fishing and seafood fraud. The Task Force is co-chaired by the designees of the Secretaries of Commerce and State. The President charged the group of senior-level federal officials with devising recommendations for a comprehensive framework of integrated programs. On December 16, 2014, the task force made recommendations to the National Ocean Council under four themes: international cooperation, enhanced enforcement tools, public-private partnerships, and a program tracing seafood from harvest to entry into commerce.
- The United Nations Food and Agriculture Organization (FAO) endorsed voluntary guidelines that address the duties and obligations of flag States in combating IUU fishing. This action by the FAO's Committee on Fisheries (COFI), in June 2014, encourages self-evaluation and reform by countries that have failed to live up to their existing responsibilities to regulate fishing vessels in their registries. RFMOs could use such assessments to evaluate countries' applications for cooperating non-member status in the organizations.
- As a result of strong U.S. leadership in 2013 and 2014, nearly all RFMOs in which the United States participates adopted measures requiring all eligible vessels to use the International Maritime Organization's (IMO) numbering scheme. IMO numbers will aid management authorities in combating IUU fishing by ensuring the accurate identity of vessels, regardless of change of name, ownership, or flag.

List of Acronyms

Acronym	Full Name
ACAP	Agreement on the Conservation of Albatrosses and Petrels
AIDCP	Agreement on the International Dolphin Conservation Program
CBP	U.S. Customs and Border Protection
CCAMLR	Commission for the Conservation of Antarctic Marine Living Resources
CDS	Catch documentation scheme
CITES	Convention on International Trade in Endangered Species of Wild Fauna and Flora
CMM	Conservation and management measure
CMS	Convention on the Conservation of Migratory Species of Wild Animals
COFI	Committee on Fisheries of the FAO
CoP16	Sixteenth Conference of the Parties (CITES)
CPC	Contracting parties and cooperating non-contracting parties, entities, or fishing entities (ICCAT)
DOS	United States Department of State
DWFDA	Distant Water Fisheries Development Act (Republic of Korea)
EC	European Commission
EEZ	Exclusive Economic Zone
EPO	Eastern Pacific Ocean
ESA	Endangered Species Act
EU	European Union
FAO	United Nations Food and Agriculture Organization
HSFCA	High Seas Fishing Compliance Act
IAC	Inter-American Convention for the Protection and Conservation of Sea Turtles
IATTC	Inter-American Tropical Tuna Commission
ICCAT	International Commission for the Conservation of Atlantic Tunas
IMCS	International Monitoring, Control and Surveillance Network
IMO	International Maritime Organization
IOTC	Indian Ocean Tuna Commission
IUU	Illegal, unreported, and unregulated (fishing)
MCS	Monitoring, control, and surveillance

MMPA	Marine Mammal Protection Act
MOU	Memorandum of understanding
MSA	Magnuson-Stevens Fishery Conservation and Management Act
MSRA	Magnuson-Stevens Fishery Conservation and Management Reauthorization Act of 2006
NAFO	Northwest Atlantic Fisheries Organization
NASCO	North Atlantic Salmon Conservation Organization
NCP	Non-Contracting Party (CCAMLR)
NEAFC	North East Atlantic Fisheries Commission
NGO	Non-governmental organization
NMFS	National Marine Fisheries Service (a NOAA line office)
NOAA	National Oceanic and Atmospheric Administration (an agency of the Department of Commerce)
NPAFC	North Pacific Anadromous Fisheries Commission
NPFC	North Pacific Fisheries Commission
PLMRs	Protected living marine resources
PSMA	Agreement on Port State Measures to Prevent, Deter, and Eliminate Illegal, Unreported and Unregulated Fishing
RFMO	Regional fisheries management organization/arrangement
SCA	Shark Conservation Act of 2010
SCRS	Standing Committee on Research and Statistics (ICCAT)
SEAFO	South East Atlantic Fisheries Organization
SPRFMO	South Pacific Regional Fisheries Management Organization
TED	Turtle excluder device
UNFSA	United Nations Fish Stocks Agreement
UNGA	United Nations General Assembly
USAID	United States Agency for International Development
USCG	United States Coast Guard
VME	Vulnerable marine ecosystem
VMS	Vessel monitoring system
WCPFC	Western and Central Pacific Fisheries Commission
WECAFC	Western Central Atlantic Fisheries Commission

I. Introduction and Background

In 2006, through Title IV of the Magnuson-Stevens Fishery Conservation and Management Reauthorization Act, Congress recognized the need for international cooperation to address some of the most significant issues affecting international fisheries today: IUU fishing and fishing practices that may undermine the sustainability of living marine resources. In 2010, Congress focused on the need for enhanced international action to conserve and protect sharks in passing the Shark Conservation Act. The statutory provisions aimed at eliciting international cooperation on these issues are codified as part of the Moratorium Protection Act, 16 U.S.C. 1826d-k.

Central to that statutory scheme is the requirement that the Secretary of Commerce, in biennial reports, identify nations whose fishing vessels are engaged in certain IUU fishing, bycatch, and shark fishing practices; describe U.S. consultations with the identified countries to urge appropriate actions; and certify whether such actions subsequent to identification have adequately addressed the offending activities. The Secretary of Commerce has delegated authority to identify and certify countries under the Moratorium Protection Act to the NOAA Assistant Administrator for Fisheries.

The Moratorium Protection Act also directs the Secretary of Commerce, in consultation with the Secretary of State and in cooperation with relevant regional fishery management councils and any relevant advisory committees, to take certain actions to improve the effectiveness of international fishery management organizations in conserving and managing stocks under their jurisdiction. These actions include urging those organizations of which the United States is a member to:

- Incorporate multilateral market-related measures against member or non-member governments whose vessels engage in IUU fishing.
- Seek adoption of lists that identify fishing vessels and vessel owners engaged in IUU fishing.
- Seek adoption of a centralized vessel monitoring system (VMS).
- Increase use of observers and technologies to monitor compliance with conservation and management measures.
- Seek adoption of stronger port State controls in all nations.
- Adopt shark conservation measures, including measures to prohibit removal of any of the fins of a shark (including the tail) and discarding the carcass of the shark at sea.
- Adopt and expand the use of market-related measures to combat IUU fishing, including import prohibitions, landing restrictions, and catch documentation schemes (CDSs).

The Secretary of Commerce is also to encourage other nations to take all steps necessary, consistent with international law, to adopt measures and policies that will prevent fish or other living marine resources harvested by vessels engaged in IUU fishing from being traded or imported into their nations or territories. The Act calls on the Secretary of Commerce, to the greatest extent possible based on availability of funds, to provide assistance to nations whose vessels are identified for IUU fishing, bycatch of PLMRs, or shark fishing practices, so that they may qualify for positive certifications.

The amended Act directs the Secretaries of Commerce and State to seek to enter into international agreements for shark conservation, including measures to prohibit removal of any fins and discarding the carcass at sea, that are comparable to U.S. measures, taking into account different conditions.

The Secretary of Commerce submitted the first Biennial Report to Congress in January 2009, the second Biennial Report in January 2011, and the third Biennial Report in January 2013. Those reports and the current one survey efforts by the United States to strengthen its leadership toward improving international fisheries management and enforcement, particularly with regard to IUU fishing, bycatch of PLMRs, and certain shark fishing practices. They also describe progress in the international arena to deal with these issues. The reports address the status of international living marine resources and contain information on actions taken to assist other countries in achieving sustainable fisheries and minimizing bycatch and discards.

As the legislation emphasizes the importance of addressing IUU fishing, PLMR bycatch, and certain shark fishing practices, the sections below provide background information on those activities and a brief discussion of other U.S. statutes and undertakings that are useful in managing U.S. fisheries responsibly and in addressing unacceptable practices in international fisheries.

A. Illegal, Unreported, and Unregulated (IUU) Fishing

The international community uses the term "IUU fishing" to describe activity that does not comply with national, regional, or global fisheries conservation and management obligations, wherever such fishing occurs. Unregulated fishing also may occur where no management authority or regulation is in place.[1]

IUU fishing activity affects fisheries of all types – from small-scale to industrial. Shipment, processing, landing, sale, and distribution of IUU fish and fish products perpetuate the financial reward from illegal harvests. IUU fishing undermines efforts of nations and RFMOs to manage fisheries in a responsible manner. It also affects the ability of governments to support sustainable livelihoods of fishermen and, more broadly, to achieve food security.

Because IUU fishing activities are generally carried out covertly, monitoring and detection are difficult. This renders quantification of the problem elusive. The FAO considers IUU fishing a serious threat to high-value fisheries that are already overfished; to marine habitats, including vulnerable marine ecosystems (VMEs); and to food security and the economies of developing countries.[2] IUU fishing activities have widespread economic and social consequences, including depriving legitimate fishermen of harvest opportunities. IUU fishing also deprives fisheries managers of information critical to accurate stock assessments. It exacerbates the problem of discards and bycatch, because vessels engaged in illegal activities are likely to engage in unsustainable fishing practices and use non-selective gear.

IUU fishing activities tend to be dynamic, adaptable, highly mobile, and increasingly sophisticated as IUU fishermen attempt to find and exploit weak links in the international fisheries regulatory system. The use of flags of convenience, as well as ports of convenience, facilitates the wide scope and extent of IUU fishing activities.

Since IUU fishing activities are complex, a broad range of governments and entities must be involved to combat them. These include flag States, coastal States, port States, market States, international and intergovernmental organizations, the fishing industry, non-governmental organizations (NGOs), financial institutions, insurers, and consumers. The United States is a member of or has substantial interests in numerous international fisheries and related agreements and organizations (see Annex 1 for a list of those most relevant to this report). U.S. involvement in international efforts to combat IUU fishing covers tools such as IUU vessel lists; stronger port State controls; improved monitoring, control, and surveillance (MCS); market-related measures to help ensure compliance; and capacity-building assistance. Progress in addressing NOAA's plan and priorities for combating IUU fishing appear in a 2013 publication entitled *Leveling the Playing Field*, available online at http://www.nmfs.noaa.gov/ia/iuu/ltpf.pdf.

B. Bycatch of Protected Living Marine Resources (PLMRs)

The bycatch of PLMRs, such as incidentally caught or entangled sea turtles, sharks, dolphins, and other marine mammals, is also a serious issue in international fisheries. Insufficiently regulated and managed bycatch of PLMRs limits the ability of the United States and other nations to conserve these resources. Fisheries bycatch can lead to injury or mortality of protected species, and can also have significant negative consequences for marine ecosystems and biodiversity. NMFS has developed a list of PLMRs, available online at http://www.nmfs.noaa.gov/ia/iuu/msra page/msra.html.

In enacting the MSRA, Congress recognized the importance of U.S. leadership in establishing international measures to end or reduce the bycatch of PLMRs. The United States is party to a number of international agreements related to the protection of living marine resources, as well as many global, regional, and bilateral fisheries agreements (see Annex 1). This report, in Part VIII, describes recent actions the United States has taken in these forums to pursue strengthened bycatch reduction measures comparable to those of the United States.

Although the statutory definition of PLMRs does not include seabirds, they are an international living marine resource for which conservation is an issue of growing global concern, and an issue in which NMFS has been actively involved internationally.[3] Section 316 of the Magnuson-Stevens Fishery Conservation and Management Act (MSA) highlights the need for the Secretary of Commerce to work cooperatively with the Secretary of the Interior and industry, and within international organizations, to seek ways to mitigate seabird bycatch. Annex 3 to this report highlights recent efforts to protect this international living marine resource.

C. Shark Conservation and Protection

Sharks are an ancient and highly diverse group of fish that present an array of issues and challenges for fisheries conservation and management due to their biological and ecological characteristics and the lack of reliable data reported on the catch of each species. Most sharks are apex predators. Many shark species are characterized by relatively slow growth, late maturity, and low reproductive rates, which can make them particularly vulnerable to overexploitation and slow to recover once stocks are depleted. As exploitation rates for some shark species and particularly the demand for fins have increased, concern has grown regarding the status of many shark stocks and the sustainability of their exploitation in world fisheries.

The United States continues to be a leader in promoting shark conservation and management globally through ongoing consultations regarding the development of international agreements consistent with the Shark Finning Prohibition Act and the SCA. The United States is committed to working bilaterally and multilaterally to promote shark conservation and management and to prevent shark finning, so that legal and sustainable fisheries are not disadvantaged by these activities. For example, within the RFMO context, the United States has focused on efforts to improve data collection for sharks, develop species-specific CMMs, promote fins-attached policies globally, and review compliance with agreed measures.

D. Other U.S. Statutes That Address IUU Fishing, PLMR Bycatch, and Shark Conservation

In addition to the Moratorium Protection Act and SCA, the United States has numerous legal tools to address IUU fishing, shark conservation, and PLMR bycatch, both domestically and internationally. These include the MSA, Lacey Act, Pelly Amendment to the Fishermen's Protective Act of 1967, Marine Mammal Protection Act (MMPA), Endangered Species Act (ESA), and International Dolphin Conservation Program Act.

Through the MSA, the United States has issued comprehensive regulations that govern all of the major fisheries in the EEZ, out to 200 miles from U.S. coasts. In the U.S. Atlantic EEZ, NMFS directly manages sharks and other highly migratory species, except for spiny dogfish (which are jointly managed by the Mid-Atlantic and New England Fishery Management Councils). In the U.S. Pacific EEZ, three regional fishery management councils – Pacific, North Pacific, and Western Pacific – are responsible for developing fishery management plans for these species. The MSA requires the Secretaries of State and Commerce to seek to secure international agreements with standards and measures for bycatch reduction comparable to those applicable to U.S. fishermen.

The SCA prohibits any person subject to U.S. jurisdiction from removing any of the fins from a shark (including the tail) at sea, having custody of a shark fin not naturally attached to the carcass, or transferring or landing any such fin.[4] In addition, it prohibits landing a shark carcass without its fins naturally attached. NMFS is developing separate rules for domestic fisheries to carry out these prohibitions.

U.S. law and policy establish a number of domestic requirements designed to reduce bycatch and other harmful effects of fishing activities on PLMRs by vessels subject to U.S. jurisdiction. U.S. fishermen are subject to requirements concerning the taking of marine mammals under the MMPA, and to rules governing fishing and related actions that affect species listed as endangered or threatened under the ESA. In addition, the MMPA requires the Secretary of Commerce, working through the Secretary of State, to develop bilateral or multilateral agreements with other nations to protect and conserve marine mammals. Section 101(a)(2) of the MMPA requires the banning of imports of fish caught with commercial fishing technology that results in the incidental kill or serious injury of ocean mammals in excess of U.S. standards.[5]

The International Dolphin Conservation Program provides that nations whose vessels fish for yellowfin tuna with purse seine nets in the Eastern Tropical Pacific are permitted to export such tuna to the United States only if the nation provides documentary evidence that it participates in the International Dolphin Conservation Program and is a member (or applicant member) of the Inter-American Tropical Tuna Commission (IATTC), is meeting its obligations under the Program and the IATTC, and does not exceed certain dolphin mortality limits.

The Pelly Amendment to the Fishermen's Protective Act provides for the possibility of trade-restrictive measures when the Secretary of Commerce certifies to the President that nationals of a foreign country are, directly or indirectly, conducting fishing operations in a manner or under circumstances that diminish the effectiveness of an international fishery conservation program, or the Secretary of Commerce or the Secretary of the Interior certifies that nationals of a foreign country are engaging in trade or taking that diminishes the effectiveness of any international program for endangered or threatened species. The President has discretion in whether to direct the Secretary of the Treasury to prohibit the importation of products from the certified country. An important enforcement tool is the Lacey Act, which prohibits interstate and foreign trafficking in fish or wildlife taken in violation of domestic or foreign law. It also prohibits the import, export, transport, sale, possession, or purchase of any fish or wildlife taken, possessed, transported, or sold in violation of any law, treaty, or regulation of the United States or in violation of any Indian tribal law. The Lacey Act provides for both civil and criminal sanctions.

Regulations under other statutes, such as the Atlantic Tunas Convention Act, ensure that U.S. fishermen are subject to the conservation measures adopted under international agreements to which the United States is a party. NMFS is currently preparing a proposed rule to revise the regulations that govern all fishing by U.S. vessels on the high seas, pursuant to the High Seas Fishing Compliance Act (HSFCA). The rule would improve administration of the HSFCA program through adjustments to permitting and reporting procedures. Issues that may be addressed in the proposed rule include monitoring of U.S. vessels operating on the high seas, placement of observers, reporting of transshipments, and bottom fishing activities. See Annex 2 for certain of these statutes, and for recent enforcement cases with an international nexus.

E. Presidential Task Forces on IUU Fishing and Wildlife Trafficking

IUU Fishing

On June 17, 2014, the White House released a Presidential Memorandum that, among other actions, established a Presidential Task Force on Combating Illegal, Unreported, and Unregulated (IUU) Fishing and Seafood Fraud and directed the Task Force to report to the President within 180 days with "recommendations for the implementation of a comprehensive framework of integrated programs to combat IUU fishing and seafood fraud that emphasizes areas of greatest need."

The Task Force is co-chaired by the designees of the Secretaries of Commerce and State. The designated co-chairs are Dr. Kathryn Sullivan (Under Secretary of Commerce for Oceans and Atmosphere and NOAA Administrator) and Catherine Novelli (Under Secretary of State for Economic Growth, Energy, and the Environment). In addition to the co-chairs, the Task Force includes designated senior-level representatives from the following agencies:

- Department of Defense
- Department of Justice
- Department of the Interior
- Department of Agriculture
- Department of Health and Human Services
- Department of Homeland Security

- Federal Trade Commission
- Office of Management and Budget
- Council on Environmental Equality
- Office of Science and Technology Policy
- Office of the United States Trade Representative
- United States Agency for International Development

The Task Force, on July 28, 2014, held a scoping session for members to discuss the complexities of the issues and means of collaboration, to share existing tools for combating IUU fishing and seafood fraud, and to identify gaps in authority. Following the initial meeting, the Task Force initiated a public engagement process in August by publishing a Federal Register notice to gain broad input and expertise from key stakeholders and interest groups to inform and advise the Task Force in developing recommendations. The Task Force held a second meeting September 29, 2014, to consider public engagement and formulate thematic interagency staff-level working groups. Five Working Groups – Entry into U.S. Commerce, Information, International Governance, Capacity Building, and U.S. Processors, Retailers, and Consumers – held numerous discussions during the month of October to develop draft recommendations for the Task Force's consideration. The Task Force met for a third time November 3, 2014, to discuss and consolidate the recommendations. On December 18, 2014, the Task Force published its recommendations in the Federal Register, solicited public comments, and later sent out a demarche to nations with marine coastlines or fishing interests, itemizing the finalized recommendations for public and international comment.

The Task Force's recommendations related to IUU fishing fall under four general themes: combating IUU fishing at the international level; strengthening enforcement and enhancing enforcement tools to combat IUU fishing; creating and expanding partnerships with state and local governments, industry, and NGOs to identify and eliminate seafood fraud and the sale of IUU seafood in U.S. commerce; and creating a risk-based traceability program to track seafood from harvest to entry into U.S. commerce.

International recommendations include working with Congress to pass implementing legislation for the Agreement on Port State Measures; developing best practices for catch documentation and MCS measures to promote adoption in RFMOs; incorporating IUU fishing threat analysis and monitoring as part of U.S. and international efforts to increase maritime domain awareness; working with U.S. trading partners to combat IUU fishing through existing and future free trade agreements and pursuing elimination of harmful fisheries subsidies; and prioritization of capacity building to address IUU fishing.

The enforcement-related recommendations include developing a strategy to optimize the collection, sharing, and analysis of information and resources to prevent IUU product from entering U.S. commerce; working with foreign customs counterparts to exchange relevant information on IUU seafood; standardizing and clarifying the rules on identifying the species, common name, and origin of seafood products across U.S. government agencies; adjusting U.S. tariff codes to enhance identification of trade in species subject to IUU fishing; and enhancing cooperation and information sharing with local and state enforcement agencies.

Noting the critical nature of private and public sector partnerships, the Task Force recommendations also include establishing a regular forum with harvesters, dealers, importers, retailers, processors, non-governmental organization (NGOs), and other groups to enhance collaboration and understanding of the levels and nature of IUU fishing.

Finally, the Task Force recommendations include creation of a risk-based traceability program to track seafood from harvest to entry into U.S. commerce. Such a program would first be applied to seafood species that are particularly vulnerable to IUU fishing, with a goal of eventually expanding the program to all seafood at first point of sale or import.

Wildlife Trafficking

Trafficking in illegally harvested wildlife, including trade in protected marine species and illegally caught fisheries products, is a multibillion-dollar business that compromises the sustainability of wild populations of the targeted species. Furthermore, wildlife trafficking undermines food security and political stability in communities dependent on these resources for their livelihoods. On July 2, 2013, President Obama issued Executive Order 13648 to more effectively mobilize government-wide efforts to combat wildlife trafficking. The order established the Presidential Task Force on Wildlife Trafficking, a high-level group co-chaired by the Departments of the Interior, State, and Justice, with participation from NOAA.

In early 2014, based on the efforts of the Wildlife Trafficking Task Force and reflecting input from its advisory committee, the President announced the National Strategy for Combating Wildlife Trafficking. The strategy, which reflects the importance of addressing marine-related issues, is based on three pillars: strengthening domestic and global enforcement; reducing demand for illegally traded wildlife at home and abroad; and strengthening partnerships with international partners, local communities, NGOs, private industry, and others to control illegal wildlife poaching and trade. Through implementation of the Strategic Plan, NOAA will continue to strengthen interagency partnerships to address illegal trade in fisheries and poaching of protected marine species, such as totoaba, sea turtles, and marine mammals.

II. Identification and Certification Provisions

In 2011, NMFS published a final rule establishing procedures to implement the identification and certification provisions of the Moratorium Protection Act for IUU fishing, bycatch of protected species, and shark catch on the high seas without a regulatory program comparable in effectiveness to that of the United States. NMFS amended these procedures, primarily to revise the definition of IUU fishing and to implement the identification and certification provisions of the SCA, through a final rule published January 16, 2013. Those procedures are described below for each type of identification.

At the beginning of the identification process, NMFS gathers information from many sources relevant to determining whether a nation's vessels have been engaged in activity that could lead to an identification. NMFS seeks public input through Federal Register notices. For this report, NMFS published, on March 5, 2014, a request for information on IUU fishing, PLMR bycatch activities, and shark fishing. NMFS then seeks corroboration or refutation from each nation under consideration, and encourages it to take action to address the activity. In deciding whether to make such an identification, NMFS considers a number of factors, which are listed in the rule.

The regulations detail the post-identification notification and consultation process, after which NMFS provides a preliminary positive or negative certification to a nation identified for having vessels engaged in IUU fishing, PLMR bycatch activities, or certain shark fishing activities. An identified nation has the opportunity to respond before the final certification is issued. The rule lists factors NMFS considers in making a final certification decision, including the effectiveness of any corrective actions taken by the identified nation.

If an identified nation takes appropriate actions, it receives a positive certification. If it receives a negative certification, sanctions under the statute may be applied under the High Seas Driftnet Fisheries Enforcement Act, including prohibitions on importation of certain fish and fish products into the United States, denial of port privileges, and other measures, under specified circumstances. The final rule describes how recommendations will be made and any sanctions implemented, in the event a nation receives a negative certification.

A. IUU Fishing

Section 609(a) of the Moratorium Protection Act requires the Secretary of Commerce to identify a nation whose vessels are engaged, or have been engaged in the preceding two years, in IUU fishing, taking into account whether the relevant international organization has failed to implement effective measures to end IUU fishing, or where no international fishery management organization with a mandate to regulate the fishing activity exists.

As Section 609(a) refers to activities of "vessels," a nation must have more than one vessel engaged in IUU activities during the "preceding two years" from the date of submission of the biennial report to Congress, to be identified for fishing in violation of international measures. Information concerning activities outside that time period cannot form the basis for an identification decision. In addition, activities during the relevant time period that are not discovered or reported before the end of the year preceding submission of the report to Congress cannot form the basis for an identification.[6]

Section 609(e)(3) of the Moratorium Protection Act requires the Secretary of Commerce to publish a regulatory definition of "illegal, unreported, or unregulated fishing," including certain minimal elements. NMFS published a definition on January 12, 2011. Subsequently, NMFS broadened and clarified the earlier regulatory definition through the January 2013 final rule (50 CFR 300.200-209). Currently the elements set out below are considered under the IUU fishing definition.

Fishing in Violation of International Measures

The first prong of the IUU fishing definition covers activities that violate measures required of a party under an international fishery management agreement to which the United States is also a party. NMFS' analytical team gathered information on incidents where RFMO compliance measures may have been violated. The team began with a search of publicly available RFMO materials, including annual reports, compliance committee meeting summaries, and IUU vessel lists. The team also searched reports from the United States Coast Guard (USCG), foreign governments, media, and NGOs.

At that point, the team began to synthesize these data: current flag, flag at listing, vessel name (current and previous), specific RFMO infractions and dates, additional infractions, and comments. Once these data were entered into a compendium, the team identified information gaps. For example, the RFMO report might not have included the IMO number for a specific vessel, or may not have identified the specific RFMO measure violated.

To fill these gaps and where possible to verify existing information, the team reviewed applicable measures and identified the specific violations that occurred. Although the team drew together the reporting and knowledge of IUU fishing activities from numerous organizations and people with years of experience in these matters, some information gaps still exist.

Based on the analysis of all available information, NMFS determined nine nations to be of interest for having unresolved allegations of violating international measures during the relevant time period (2013 and 2014), and therefore considered them for potential identification.

Through diplomatic channels, NMFS contacted these nations to verify information regarding alleged IUU fishing activities by their vessels, as well as possible actions or inactions by the nations themselves in violation of international measures. From the responses of five of the nine countries and from other sources, NMFS collected information that either refuted the allegations or showed that corrective actions had been taken to address the IUU fishing activities of concern. Information provided by the remaining four nations failed to demonstrate that sufficient corrective action had been taken to address all of the potential violations.

In a case where action taken by a nation was pending against a vessel, but no resolution had been reached to exonerate or sanction the vessel, NMFS considered the activities of the vessel to be a foundation for identification. NMFS also considered the activities of a vessel as a basis for identification when the agency was unable to ascertain why a case against the vessel suspected of IUU fishing had been closed.

Detailed information on the four countries identified for this type of IUU fishing appears below in Part III.A; information on countries of interest that were not identified is found in Part III.B.

Undermining RFMO Conservation by Non-Parties

Under this aspect of the IUU fishing definition, a nation may be identified for fishing activities that undermine the conservation of resources under an international fishery management agreement to which the United States is a party, despite the fact that the nation is not a party to the agreement. NMFS has identified Nigeria for undermining conservation measures of an RFMO of which it is not a member.

Overfishing of Shared Stocks

This element of the definition of IUU fishing includes overfishing of stocks shared by the United States in areas without international measures or management organizations. As of June 30, 2014, NMFS has assessed the following five highly migratory stocks as both overfished and shared by U.S. and foreign fleets: Atlantic blue marlin, western Atlantic bluefin tuna, Atlantic white marlin, Pacific bluefin tuna, and central western Pacific striped marlin. The first three stocks are managed by an international management organization, the International Commission for the Conservation of Atlantic Tunas (ICCAT). The Western and Central Pacific Fisheries Commission (WCPFC) and the IATTC have adopted specific management rules for Pacific bluefin tuna. The WCPFC has management measures for central western Pacific striped marlin, but the IATTC does not. NMFS is not identifying any country as conducting this type of IUU fishing for these particular stocks in 2013–2014. Other shared stocks assessed as overfished were found aboard vessels that were illegally fishing in U.S. waters in the Gulf of Mexico. Those species include red snapper, gag grouper, and gray triggerfish. NMFS has identified Mexico for overfishing of shared stocks in this report.

Destructive Fishing Practices on VMEs

This part of the definition includes fishing activity that has a significant adverse impact on VMEs, including seamounts, hydrothermal vents, and cold water corals, located in areas beyond national jurisdiction, for which there are no applicable conservation or management measures or in areas with no applicable international fishery management organization or agreement. During the reporting period, NMFS found no nation that had conducted IUU fishing activities under this prong of the IUU fishing definition. Currently five RFMOs actively manage bottom fishing: the Northwest Atlantic Fisheries Organization (NAFO), the North East Atlantic Fisheries Commission (NEAFC), the South East Atlantic Fisheries Organization (SEAFO), the Convention for the Conservation of Antarctic Marine Living Resources

(CCAMLR), and the South Pacific Regional Fisheries Management Organization (SPRFMO). Each of these organizations, as well as one in formation (the North Pacific Fisheries Commission, NPFC), have measures to protect VMEs from bottom fishing activities, in accordance with the 2006 United Nations General Assembly (UNGA) Sustainable Fisheries Resolution (61/105) and reflecting guidance from the FAO's International Guidelines for Deep Sea Fisheries.[7] Nations fishing in accordance with the rules of these organizations, by definition, would not meet the criteria for IUU fishing identification under the Moratorium Protection Act.

The Southwest Atlantic Ocean, a small portion of the Pacific Ocean,[8] and the Indian Ocean are the only areas of the high seas where bottom fishing is not being managed under an RFMO.[9] To avoid identification under the Moratorium Protection Act, States with vessels known to be fishing in these areas in 2013 and 2014 must have had measures in place to prevent significant adverse impacts to known or likely VMEs.

In response to the 2009 UNGA review of Resolution 61/105, the FAO maintains a list of vessels authorized for bottom fishing on the high seas. In 2013 and 2014, only Japan reported having authorized vessels to bottom fish in areas outside of RFMO convention areas, although several States, including Australia, the Cook Islands, New Zealand, the Republic of Korea, the Russian Federation, Spain, and the United Kingdom, had previously reported having authorized vessels for bottom fishing on the high seas. Those nations informed the UN, and confirmed through consultations with NMFS, that all fishing activities were being conducted in accordance with Resolution 61/105. NMFS therefore concludes that they would not qualify as IUU fishing.

NMFS will continue to work with international partners to strengthen implementation of and compliance with existing RFMO management measures. NMFS will also continue to support scientific research to identify VMEs on the high seas and gear modifications to reduce the impact of bottom-tending gears on vulnerable habitats.

Foreign Fishing in U.S. Waters

This new aspect of the IUU fishing definition allows for identification of a nation when its vessels fish without authorization in the U.S. EEZ. NMFS has identified Mexico for fishing in the U.S. EEZ in this report.

B. Bycatch of PLMRs

Section 610(a)(1) of the Moratorium Protection Act requires the Secretary of Commerce to identify a nation for bycatch activities if:

- fishing vessels of that nation are engaged, or have been engaged during the preceding calendar year, in fishing activities or practices in waters beyond any national jurisdiction that result in bycatch of a protected living marine resource, or beyond the exclusive economic zone of the United States that result in bycatch of a protected living marine resource shared by the United States;
- the relevant international organization for the conservation and protection of such resources or the relevant international or regional fishery organization has failed to implement effective measures to end or reduce such bycatch, or the nation is not a party to, or does not maintain cooperating status with, such organization; and
- the nation has not adopted a regulatory program governing such fishing practices designed to end or reduce such bycatch that is comparable to that of the United States, taking into account different conditions.

The first criterion of Section 610(a) contains a shorter time frame for identification than the comparable provision for IUU fishing in Section 609(a) – just the preceding calendar year – and shares the restriction that an identification cannot be based on the activities of a single vessel.[10]

The regulations define “bycatch” to mean “the incidental or discarded catch of protected living marine resources or entanglement of such resources with fishing gear” (50 CFR 300.201). For purposes of the Moratorium Protection Act (Section 610(e)), the term “PLMR”:

- includes non-target fish, sea turtles, or marine mammals that are protected under U.S. law or international agreement, including the MMPA, ESA, Shark Finning Prohibition Act, and CITES, but
- does not include species, except sharks, managed under the MSA, the Atlantic Tunas Convention Act, or any international fishery management agreement.

Over the past two years, NMFS collected significant amounts of information on activities from numerous sources, including government and academic studies, relevant international organizations, NGOs, and the media. NMFS' team of subject matter experts examined the bycatch in question and any relevant regulations or management measures. Based on analysis of all available information, NMFS determined no nation to be of interest for having vessels that allegedly engaged in PLMR bycatch, and therefore considered none for identification.

The failure to identify a country for PLMR bycatch activities during this process is due to limitations within the Moratorium Protection Act rather than significant reduction in PLMR bycatch in global fisheries. For example, a number of nations self-report bycatch of PLMRs to RFMOs or other international organizations, but those data are not generally available in time for action under the Act. Many countries publish bycatch reports and corresponding analyses in the year after the data are collected, or even later. International organizations and journal publications often report these data several years after they receive the information.

During the next two years, NMFS will continue to collect information for possible identification of nations for PLMR bycatch under the provisions of the Moratorium Protection Act. To support this work, NMFS will collaborate with international partners to improve reporting and collection of bycatch incidents within relevant international conservation organizations. While some RFMOs collect bycatch information, reporting is often voluntary. Most RFMOs that do collect data have not standardized their data collection. Those that do have standards often receive data from nations that do not use these standardized formats, which creates significant gaps in the technical information available and reduces the ability of these organizations to address bycatch issues.

NMFS will continue to be a leader bilaterally, multilaterally, and globally to reduce bycatch of PLMRs. NMFS has long-standing outreach and assistance programs with a number of nations, through cooperative research or other capacity-building activities, to reduce and mitigate bycatch (see Part X for examples). NMFS intends to continue to support existing capacity-building efforts, where appropriate, and to initiate additional programs with other nations based on the nature of their PLMR bycatch interactions, need for assistance, and willingness to work cooperatively with the United States. NMFS will also continue to promote comprehensive CMMs through international organizations to reduce bycatch of PLMRs, by working with international partners to improve assessment of the impact of fisheries on bycatch taxa, support research into gear modifications and alternative gear types, and develop management measures to reduce bycatch.

C. Shark Conservation and Protection

In the 2006 legislation, sharks fell implicitly within the definition of "IUU fishing" and explicitly within the PLMR definition. As mentioned above, the SCA amended the guidelines for defining IUU fishing to specify that violation of shark conservation measures is included. The SCA requires the Secretary of Commerce to identify nations whose vessels are engaged, or have been engaged during the preceding calendar year, in fishing activities or practices on the high seas that target or incidentally catch sharks, and the nation has not adopted a regulatory program for the conservation of sharks, including measures to prohibit removal of any of the fins of a shark (including the tail) and discarding the carcass of the shark at sea, that is comparable to that of the United States, taking into account different conditions.[11]

Identification of nations under the provisions of the SCA may be based only on activities occurring on the high seas during the calendar year preceding submission of the biennial report to Congress; thus for the 2015 report the activities must have occurred during 2014.[12] During the past two years, NMFS analyzed information from the websites of many international organizations: the FAO, ICCAT, the IATTC, the WCPFC, the Indian Ocean Tuna Commission (IOTC), NAFO, the General Fisheries Commission for the Mediterranean, SEAFO, and CCAMLR. A number of nations reported catching sharks, but none of the activities met the SCA criteria because they took place prior to 2014. Normally, nations report the prior year's catch to RFMOs. For example, at the 2014 annual meeting of ICCAT, the catch reported by members was for 2013. A further complicating factor is that the location of the catch of sharks is not reported; one cannot discern whether the catch occurred on the high seas or within EEZs. Therefore, NMFS does not have any applicable data for shark catch on the high seas and is not identifying any nation under Section 610(a)(2) of the Moratorium Protection Act.

III. Identifications under Section 609

Violations of IATTC resolutions resulted in the identification of three nations that are members of that organization: Colombia, Ecuador, and Nicaragua. Portugal is identified for its vessels' violations of the conservation measures of NAFO, of which it is a member. NMFS is identifying Nigeria for fishing activities that undermine the effectiveness of CCAMLR, of which is it not a member. Mexico is identified for two types of IUU fishing activities: overfishing of shared stocks and illegal fishing in the U.S. EEZ.

A. Nations Identified

1. Colombia

Bases for Identification

Colombia is being identified because four Colombian-flagged vessels engaged in fishing activities that violated IATTC conservation measures.[13] Records from the IATTC indicate a number of Colombia's fishing vessels violated IATTC resolutions, including C-11-03 (prohibiting fishing on data buoys), C-12-01 (on tuna conservation), and C-04-05 Rev 2 (on bycatch). These violations include interacting with a data buoy, making transits during a closure period without a permit exempting the vessel from carrying an observer, and the discarding of plastic bags or salt bags overboard.

Evidence from NOAA's National Data Buoy Center indicates the fishing vessel *Grenadier* interacted with a data buoy, in violation of IATTC Resolution C-11-03. On June 4, 2014, the fishing vessel was identified via Automatic Identification System data near TAO buoy 2S-95W and was photographed by a camera mounted on the buoy. The data show a fishing vessel approaching the buoy and departing shortly thereafter. Photos show a fishing vessel approaching the data buoy, launching a speed boat, and briefly interacting with the data buoy before departing.

Fishing vessels *Patricia Lynn* and *Dominador* made transits during a closure period without having a permit exempting them from carrying an observer, violating IATTC Resolution C-12- 01. The *Patricia Lynn* transited from Manta to Guayaquil on July 31, 2013, and then from Guayaquil to Manta from August 18, 2013, to August 20, 2013. The *Dominador* transited between the ports of Flamenco (Panama) and Manta (Ecuador) in November 2013. The Secretariat did not receive notice that the vessels had permits exempting them from observer coverage requirements during these transits, which were made during a closure period. NMFS understands that Colombia is investigating these transits.

The fishing vessel *Amanda S* discarded plastic bags or salt bags in July 2013, violating IATTC Resolution C-04-05 Rev 2, which requires that contracting parties prohibit their vessels targeting species covered by the Convention from disposing of salt bags or any other type of plastic trash at sea, in order to prevent sea turtle mortality. NMFS understands that Colombia is investigating this activity.

Outreach to Colombia

NMFS sent a letter to Colombia dated October 15, 2014, and received a response letter dated November 17, 2014, with follow-up information provided in a letter dated December 10, 2014. In its response, Colombia acknowledged the reported cases, stated preliminary inquiries are underway, and noted that any new information would be provided as soon as it was available. Regarding the *Grenadier*'s data buoy interaction, Colombia explained it lacks domestic regulations governing interactions between vessels and data buoys. The United States will follow up with Colombia on this matter during the consultation.

2. Ecuador

Bases for Identification

Ecuador is being identified based on IATTC records showing that a number of Ecuadorian-flagged vessels violated IATTC resolutions in 2013 or 2014.[14] Ecuador provided information indicating that many of these vessels are still under investigation, or their sanctions are under appeal.

The *El Conde* and *Mariella* fished near data buoys in violation of IATTC Resolution C-11-03. Ecuador noted these vessels are being dealt with specifically and the results will be provided when available.

The *Mariajose* was fined $1,700 for tuna discards. According to Ecuador, 16 metric tons were discarded due to small size not suitable for market. The fine is under appeal.

The following vessels discarded plastic bags or salt bags, violating IATTC Resolution C-04-05 Rev 2, which requires contracting parties to prohibit their vessels targeting species covered by the Convention from disposing of salt bags or any other type of plastic trash at sea in order to prevent sea turtle mortality: the *Malula*, *Adriana*, *Don Mario*, *Alina*, and *Lizi* (two trips). The *Lizi* received fines of $2,380 and $1,700; both fines are currently under appeal and may involve issues other than plastic discards (tuna discards and improper turtle rescue).

The following vessels carried out improper turtle rescues, violating IATTC Resolution C-04-05 Rev 2: the *Elizabeth F*, *Don Mario*, and *Chiara*.

Ecuador investigated the *Yelisava* for tuna discards and determined there was no violation of C12-01 since the discarded tuna was used as bait. That resolution, however, does not indicate "use of bait" as an acceptable discard justification, but rather requires all purse-seine vessels to first retain on board and then land all bigeye, skipjack, and yellowfin tuna caught, except fish considered unfit for human consumption for reasons other than size.

NMFS is concerned that Ecuador's short statute of limitations, 60 days from the date of violation, prevents the effective enforcement of domestic laws implementing RFMO measures. Under the Ecuadorian constitution, when a statute does not provide a statute of limitations, a default limitation of 60 days applies. This means the government has only 60 days from the date of violation in which to bring an action. Since it is common practice for RFMO Secretariats to compile and review compliance issues annually for the preceding year, information received regarding Ecuador's vessels would almost always be provided long after the 60-day statute of limitations has expired. Ecuador has reported that it is currently working to draft new fisheries legislation to address this issue. NMFS also understands that the Government of Ecuador recently reached an agreement with the IATTC Secretariat to expedite its receipt of observer reports in an effort to identify violations within the 60-day period. NMFS will discuss the progress of this information-sharing arrangement during the consultation with Ecuador.

Other Information and Fishing Activities that Did Not Form the Bases of Identification

Ecuador provided information on the following investigations, which, as appropriate, resulted in sanctions; these cases were not appealed. Ecuador investigated the allegations against the *Julia D* and the *Alize* that no effort was made to prevent turtle entanglement and determined they could not be proven. Ecuador stated that low quantities of tuna discards (0.03 tons of discarded skipjack on the *Alize*) are likely smashed product that would contaminate the rest of the catch if stored in the hold, and therefore would not be a violation of C-12-01. Ecuador also investigated the allegation that this vessel discarded five tons of skipjack and determined it could not be proven.

Ecuador fined the *Giulietta* $2,040 for violations of discarded plastic bags, shark mutilation, and use of sealed holds. Ecuador investigated the allegation against this vessel that a turtle was released improperly and determined no violation took place.

Ecuador fined the *Cap. Dany B* $1,360 for discarding plastic bags and salt bags. Ecuador stated that small quantities of tuna discards (such as those that occurred during a trip where five sets had discards of less than 0.4 tons each) are likely smashed product that would contaminate the rest of the catch if stored in the hold.

Ecuador fined the *Ciudad de Portoviejo* $3,180 for violations of discards of plastic bags or bags of salt and three tons of skipjack. Ecuador fined the *Ginno D* $1,360 for plastic trash discards. Ecuador fined the *Ugavi* $2,380 for retaining oceanic whitetip sharks. Ecuador fined the *Don Alvaro* $3,400 for fishing operations lasting more than 30 days during the second closure period and for making a trip during a closed season without notifying authorities of its permit exempting it from carrying an observer. Ecuador also fined the *Aldo* $3,400 for searching for fish during the high seas closure period. The *Maria Fatima* was fined $1,020 for transiting during a closure period without a waiver.

Ecuador investigated the tuna discards of the *Lucia T* and does not consider them to be a violation of C-12-01 based on the percentage discarded. Ecuador investigated the *Monteneme* for three tons of skipjack discards and stated there was lack of evidence to determine that an infraction occurred. Small quantities of tuna discards by the *Monteneme* (such as those that occurred during a trip with one ton of skipjack discarded) are likely smashed product that would contaminate the rest of the catch if stored in the hold.

Ecuador determined that the crew of the *Malula* made no effort to prevent sea turtle entanglement and fined the vessel $1,700. Ecuador states low quantities of tuna discards by this vessel (0.5 tons of discarded skipjack) are likely smashed product that would contaminate the rest of the catch if stored in the hold.

Ecuador fined the *Adriana* $2,040 for the disposal of salt bags at sea; it investigated the allegation of improperly rescuing a turtle and determined there was no infraction. Small quantities of tuna discards by the *Don Mario* (one ton of discarded skipjack and one ton of discarded unspecified tuna) were likely smashed product that would contaminate the rest of the catch if stored in the hold.

Ecuador fined the *Mariella* $1,700 for discarding plastic and salt bags of trash, and fined the vessel another $1,700 for discarding two plastic bags of trash during another trip. Ecuador investigated the allegation of plastic trash discards by the *Alina* and determined there was not enough evidence to fine the vessel.

Outreach to Ecuador

NMFS sent a letter to Ecuador dated October 2, 2014, and received a response letter dated October 25, 2014, with information on the status on the vessels cited in the NMFS letter.

3. *Mexico*

Bases for Identification

Mexico is being identified for having vessels fishing illegally in the U.S. EEZ, and for overfishing of stocks shared with the United States, in areas without applicable international measures or management organizations, that has adverse impacts on such stocks.[15]

While foreign fishing without authorization in the U.S. EEZ has been illegal since 1977, the definition of IUU fishing under the Moratorium Protection Act was revised only recently to include such activity.[16] The United States and Mexico have worked to address unauthorized fishing by Mexican vessels in the U.S. EEZ of the Gulf of Mexico over many years, but more violations of this type occurred in 2013.

The USCG apprehended 24 open-hulled vessels powered by outboard motors (known locally as lanchas) in the U.S. EEZ with 82 Mexican nationals onboard, along with evidence of fishing activity. The vessels had made incursions as far as 41 nautical miles into the U.S. EEZ. The USCG documented a total of 1,418 red snapper, five gag grouper, and four gray triggerfish onboard the lanchas; these three species are from stocks shared with the United States and that have been determined to be overfished by the United States. Under the Moratorium Protection Act, the definition of IUU fishing includes overfishing of stocks shared by the United States that has adverse impacts on such stocks. Mexico reported to NMFS that it considers Red Snapper to be fully exploited. NMFS is concerned about the adverse impacts of this fishing activity, particularly on Red Snapper, given the status of that stock and the large amount of catch documented.

Outreach to Mexico

NMFS sent a letter to Mexico dated October 17, 2014, and received a response letter dated December 10, 2014. In its response, Mexico described a number of initiatives it hopes will address the issue of lancha incursions into the U.S. EEZ. These include carrying out a verification of vessel registrations and permits by the end of December 2014, expediting the installation of Automatic Identification System transmitters on registered vessels, strengthening the Mexican Navy's surveillance and patrol efforts in the area, and analyzing the feasibility of establishing a fishing exclusion zone near the maritime boundary.

4. Nicaragua

Bases for Identification

Nicaragua is being identified because four Nicaraguan-flagged vessels violated IATTC conservation measures in 2013.[17]

The *Olivia* improperly handled a turtle entanglement, violating IATTC Resolution C-04-05 Rev 2 (on bycatch). NMFS understands that Nicaragua opened an administrative process regarding this case.

The *Lucile* made fishing sets near a data buoy on August 18, 2013, and on September 15, 2013, violating IATTC Resolution C-11-03 (prohibiting fishing on data buoys). NMFS understands that Nicaragua has opened an administrative process regarding these activities.

The following vessels discarded tuna in violation of IATTC Resolution C-12-01, which requires all purse seine vessels to first retain on board and then land all bigeye, skipjack, and yellowfin tuna caught, except fish considered unfit for human consumption for reasons other than size. The *Olivia* discarded tuna on August 7, 2013: two tons of yellowfin and two tons of skipjack. The *Lucile* discarded 0.1 tons of skipjack on September 20, 2013, and on September 27, 2013, discarded 0.01 tons of yellowfin and 0.01 tons of skipjack. The *Emilio* discarded 0.5 tons of skipjack on August 16, 2013. The *Capt. Joe Jorge* discarded 2.5 tons of yellowfin on November 11, 2013. NMFS understands that Nicaragua has opened an administrative process regarding these charges.

Outreach to Nicaragua

NMFS sent a letter to Nicaragua dated October 7, 2014, and received a response letter dated November 7, 2014. In its response, Nicaragua indicated all four vessels in question are still under investigation with official decisions expected to be published soon.

5. Nigeria

Bases for Identification

Nigeria is being identified because at least two of its vessels undermined the effectiveness of CCAMLR conservation measures in 2013 and 2014.[18]

Four fishing vessels listed on the CCAMLR Non-Contracting Party (NCP) IUU Vessel List were flagged by Nigeria. Information obtained by the United States suggests that two of these vessels engaged in IUU fishing in 2013 and 2014. The other two vessels have been observed in the past inside the Convention Area and have a long history of association with IUU fishing inside the Convention Area.

The *Thunder,* placed on the CCAMLR NCP-IUU Vessel List in 2006, was observed inside Statistical Division 58.4.3b of the Convention Area on February 13, 2013 (externally marked with homeport Lagos). On August 23, 2013, this vessel was sighted by a French surveillance team fishing in Statistical Subarea 58.6. An NGO reported seeing the vessel fishing inside the Convention Area on December 16, 2013.[19] In addition, the vessel was sighted several times outside the Convention Area (March 20, 2013; April 14 and 18, 2013; September 18, 2013; and April 14, 2014). On April 20, 2013, the vessel was inspected in the port of Benoa, Bali, Indonesia. During the inspection, the master of the vessel produced a certificate of registration indicating that the vessel is flagged to Nigeria. Press reports, dated April 30, 2014, indicate the vessel (using the name *Ming 5*) was detained by Malaysian authorities. According to the press report, 100 to 150 tons of fish were found onboard this and another vessel (CCAMLR IUU vessel *Chang Bai*, using the name *Tai San*) and were believed to have been caught in violation of CCAMLR conservation measures. Australian authorities report asking Nigeria to clarify this vessel's registration and receiving confirmation that the vessel is registered to Nigeria. Australian authorities note that this vessel has a long history of association with IUU fishing. It was the subject of an INTERPOL Purple Notice published December 5, 2013, at the request of New Zealand.

The *Viking*, placed on the CCAMLR list in 2004, was observed inside Statistical Division 48.6 of the Convention Area on January 24, 2014, reportedly fishing with gillnets. New Zealand and Australian authorities observed the vessel (externally marked with homeport Lagos) on two occasions outside the Convention Area (March 16, 2014; March 21, 2014). Australia obtained information that the vessel was observed docked in Colombo, Sri Lanka, on October 24, 2013. This vessel was also the subject of an INTERPOL Purple Notice published September 6, 2013, at the request of Norway. Australian authorities note that this vessel has a long history of association with IUU fishing.

Other Information and Fishing Activities that Did Not Form the Bases of Identification

While the following vessels were not sighted in the Convention Area and did not form the basis for identification, their continued sightings along routes commonly used to transit to and from the fishing grounds are of concern to the United States.

The *Lana* (homeport Lagos), placed on the CCAMLR NCP-IUU Vessel List in 2007, was observed by Australian authorities on three occasions (September 3, 2013; November 19, 2013; February 14, 2014).

The *Perlon* (homeport Lagos), placed on the CCAMLR NCP-IUU Vessel List in 2003, was observed by Australian authorities on two occasions (July 20 and 21, 2014) with external markings indicating it was flagged to Nigeria. Australian authorities note that, while these sightings were outside the Convention Area, these vessels have a long history of association with IUU fishing inside the Area, and continue to undermine conservation measures established by CCAMLR.

Outreach to Nigeria

NMFS sent a letter to Nigeria dated October 7, 2014, and received a response dated December 11, 2014, from Nigeria's Department of Fisheries. In its response, Nigeria confirmed the fishing vessels *Thunder*, *Lana*, and *Perlon* are all registered by the Ministry of Transportation. Nigeria states it has no records for the fishing vessel *Octopus*; however, NMFS notes this vessel is currently operating under the name *Viking*.

The Department of Fisheries stated that it has not licensed any of these vessels to fish in Nigerian waters or elsewhere. The Department of Fisheries stated it was not in a position to know whether the *Thunder* or *Viking* were in the Convention Area at any time, and cannot confirm that the vessels were engaged in fishing for CCAMLR species or in activities in support of other vessels fishing for CCAMLR species. The Department of Fisheries also explained that it is not currently in a position to confirm whether the Ministry of Transportation consults RFMO IUU vessel lists prior to registering vessels.

6. Portugal

Bases for Identification

Portugal is being identified because three Portuguese-flagged vessels engaged in fishing activities that violated NAFO conservation measures in 2013 and 2014.[20]

The *Calvao* had misrecorded catch in its fishing logbook and daily catch report, violating Article 28 of the NAFO Conservation and Enforcement Measures. On April 13, 2014, Canadian NAFO inspectors boarded the Portuguese fishing vessel and observed that the logbook information did not correspond with Canadian air surveillance flight catch estimates and the daily catch report for April 7, 2014. Canadian NAFO inspectors issued two citations to the master of the fishing vessel for these discrepancies.

A Portuguese fishing vessel (referred to as "CallSign 1" in the source document) had occurrences of misrecorded catch, mislabeling product, and tampering with seals, violating Articles 28.1, 38.1, 27.1, and 38.1(n) of the NAFO Conservation and Enforcement Measures. These violations were documented during a port inspection in Cangas do Morrazo, Spain, on April 12, 2013. Canadian DFO reporting indicates the vessel could be the *Santa Isabel*.

The *Coimbra* had violations of mesh sizes, violating Chapter I Article 13.2(d) and Annex III.B.2 of the NAFO Conservation and Enforcement Measures. The vessel was inspected at sea on April 19, 2014. Canadian NAFO inspectors issued the following two serious citations to the master of the fishing vessel: fishing for groundfish with a mesh size less than 130 millimeters, and use of a multiple flap topside chafer attached to the upper side of the cod end with a mesh size less than that of the cod end.

Outreach to Portugal

NMFS sent a letter to Portugal dated October 7, 2014, and received a response letter dated November 7, 2014. In its response, Portugal acknowledged the cases relating to the *Calvao* and *Coimbra* and stated that investigations are ongoing. Portugal was unaware of the concerns raised during the port inspection in Cangas do Morrazo, Spain, on April 12, 2013, and stated it would inquire further. Portugal indicated it will present the results of the investigation at the next NAFO Standing Committee on International Control in 2015, via the European Union (EU).

B. Countries "of Interest" Not Identified

1. Belize

NMFS contacted Belize because two of its registered vessels reportedly engaged in fishing activities that violated IATTC conservation measures in 2013.[21]

A false transshipment declaration was reported aboard the carrier vessel *Heng Vu*, allegedly in violation of IATTC Resolution C-12-07, which establishes a program for transshipment by large-scale fishing vessels. The transshipment recorded in the declaration occurred on December 13, 2013, in the Eastern Pacific, with 36,120 kilograms of fish transshipped by the Chinese-flagged longline fishing vessel *Xin Shi Ji 75*. China's investigation, however, concluded that vessel did not make an at-sea transshipment to the *Heng Vu* on that date.

Evidence from NOAA's National Data Buoy Center indicates the fishing vessel *Hung Chi Fu 35* violated IATTC Resolution C-11-03, which prohibits fishing on data buoys. On July 27, 2013, a fishing vessel with that name painted on its superstructure was photographed by a camera mounted on the NOAA data buoy. At the time the picture was taken, compass data from the buoy indicated it was in contact with an external force. Further analysis indicates the data buoy sustained structural damage, likely as a result of the interaction it had with the fishing vessel observed in the photograph. The data buoy became non-operational shortly afterwards. On July 9, 2014, Belize assured members of the U.S. delegation to the IATTC that it would investigate this allegation and report back. Initial information from the Belize IATTC delegation confirmed this vessel was in the vicinity at the time of the incident, but the vessel has since re-flagged to Costa Rica.

Belize stated it investigated the false transshipment allegation and determined the transshipment did not take place, so there was no need to sanction the vessel. Belize also described its efforts to investigate the *Hung Chi Fu 35*'s data buoy interaction, which included sending a notice to the vessel owner via the agent, and attempting to contact Costa Rican authorities so that appropriate action could be taken. Belize states it has national legislation in place to implement the IATTC resolution regarding data buoys, and notes the challenge of investigating a vessel that is no longer under its jurisdiction. Since Belize determined no violation took place regarding the false transshipment allegation, and has only has one vessel with an unresolved allegation, Belize is not being identified.

2. Costa Rica

Records from the USCG indicate that vessels flagged to Costa Rica may have violated an IATTC measure in 2013.[22] The fishing vessels *Bella Canan II, Triple Suerte III,* and *Vikingo II* were observed with IATTC-managed species onboard, but were not included in the IATTC regional vessel register at the time, allegedly in violation of IATTC Resolution C-11-05. That resolution requires contracting parties to prohibit fishing for, retaining on board, and transshipment and landing of, tuna and tuna-like species by longline vessels over 24 meters if they are not included in the IATTC list for such vessels. These vessels were boarded and inspected by USCG personnel under the authority of a United States-Costa Rica enforcement agreement. The *Bella Canan II* was boarded on November 27, 2013, approximately 150 nautical miles southwest of the Azuero Peninsula, Panama, and found with marlin, dorado, and sharks onboard. The *Triple Suerte III* was boarded on August 8, 2013, approximately 60 nautical miles southeast of Isla de Coiba, Panama, and found with yellowfin tuna onboard. The *Vikingo II* was boarded in March 2013 in the Eastern Pacific and found with yellowfin tuna onboard.

Costa Rica provided copies of the fishing licenses for the three vessels in question, noting they are all less than 24 meters. Costa Rica also indicated this licensing information was sent to the IATTC with the hope that by early 2015 the current register could be updated. Since Costa Rica provided documentation proving these vessels were not subject to Resolution C-11-05, and provided updated registry information to the IATTC, Costa Rica is not being identified.

3. Ghana

NMFS contacted Ghana about allegations relating to Ghana's implementation of VMS requirements under ICCAT.[23] In the ICCAT report for biennial period 2012–2013 Part II (2013) – Vol. 1, the EU states the VMS system purchased for Ghana, through an ICCAT project for the improvement of data, may not allow Ghana to fully implement ICCAT VMS requirements (Recommendation 03-14). A report from the International MCS (IMCS) Network, dated October 2013, describes Ghana's VMS as "defunct" and notes "the lack of an operational VMS system is a severe handicap to effective operations; while this problem is being addressed to some extent. . . , it will be some time before it is possible to assess the effectiveness of the system and its long term reliability and availability."

Ghana provided a number of details indicating that Ghana's VMS exceeds the minimum standard required by ICCAT, thanks in large part to a World Bank project. Namely, Ghana's Fisheries Monitoring, Control, Surveillance and Enforcement Unit became functional in October 2013 with a new staff of 61, including Navy and Marine Police. This staff received training by the EC's Directorate-General for Maritime Affairs and Fisheries on the use of VMS and other MCS tools to detect IUU fishing. In addition, an electric generator was purchased and installed to serve as an alternative power source.

NMFS also understands that Ghana plans to strengthen its VMS by increasing the frequency of data transmissions; making fishing without a functioning transponder a violation of the fishing license; and taking corrective action sooner by strengthening the monitoring of remote sensing data. As well, Ghana in early 2015 intends to introduce a license condition that vessel captains transmit daily reports to the Ghanaian Fisheries Monitoring Unit on vessel location, activities, and catch by species. Failure to report, or reporting of incorrect information, would violate the license conditions and could result in sanctions.

Since Ghana provided information indicating it has addressed allegations from the 2013 reports, and is now meeting the minimum ICCAT standards for a VMS, Ghana is not being identified.

4. Guatemala

USCG records indicate that vessels flagged to Guatemala may have violated an IATTC measure in 2013.[24] The fishing vessel *Capitan Chato* was observed with IATTC-managed species onboard; the *Maria Jose del Mar* was observed with sharks onboard. These vessels were not found in the IATTC regional vessel register at the time, allegedly in violation of IATTC Resolution C-11-05. These vessels were boarded and inspected by USCG personnel under the authority of a U.S.-Guatemala enforcement agreement. The *Capitan Chato* was boarded on October 30, 2013, approximately 120 nautical miles south of the Guatemala/Mexico border, and found with dorado and yellowfin tuna onboard. The *Maria Jose del Mar* was boarded on November 24, 2013, approximately 120 nautical miles south of the Azuero Peninsula, Panama, and found with sharks in the nets onboard. Sharks are captured in fisheries targeting tunas and tuna-like species in the IATTC Convention Area.

Guatemala provided registration information on the *Capitan Chato* indicating the vessel is less than 24 meters. It did not have any record of the *Maria Jose del Mar* and determined it is not a Guatemalan-flagged vessel. Since Guatemala provided documentation proving the *Capitan Chato* it not subject to Resolution C-11-05, and explaining how the *Maria Jose del Mar* could not be flagged to Guatemala, Guatemala is not being identified.

5. Spain

Records from NAFO indicate that vessels flagged to Spain may have violated NAFO measures in 2013 or 2014.[25] The *Puente Sabaris* reportedly had violations relating to bycatch limits, catch reporting, and stowage plans, allegedly in violation of Chapter I, Article 6.2(a), Article 28.5, Article

28.2(a)(b), and Article 38.1(i) of the NAFO Conservation and Enforcement Measures. On March 1, 2013, Canadian NAFO inspectors boarded the vessel while it was fishing for Greenland halibut in NAFO Division 3N. A post-inspection analysis of the tow-by-tow data indicated that on February 28, 2013, the master had exceeded the bycatch limit of witch flounder and had failed to move the minimum 10 nautical miles, required under the NAFO Conservation and Enforcement Measures. As a result of these findings, a subsequent inspection was conducted on March 4, 2013, and the master was issued a citation under Chapter I, Article 6.2(a). The report of violation was provided to the contracting party (the EU) for the vessel for follow-up.

On May 3, 2014, an inspection report noted that this vessel's master failed to maintain a stowage plan for freezer hold number 2 and misreported catch in the fishing logbook subject to Article 38.1(i). Found hidden under cardboard cartons in this hold were 51.25 tons of Greenland halibut product. An EU inspector made an independent inspection and confirmed that hold number 2 had that amount of undeclared Greenland halibut.

A Spanish fishing vessel (referred to as "CallSign 8" in the source document) reportedly violated the NAFO Conservation and Enforcement Measures relating to product labeling, capacity plans, and bycatch. The vessel was inspected in port in Vigo, Spain, on February 12, 2013. Apparent violations related to product labeling (Article 27), capacity plans (Article 25.9), and bycatch (Article 6.2(a)) were noted. The vessel was inspected in Vigo again on July 2, 2013. An apparent violation related to capacity plans was noted (Article 25.10(b)).

Finally, a Spanish fishing vessel (referred to as "CallSign 6" in the source document) reportedly violated the NAFO Conservation and Enforcement Measures relating to catch recording (Article 38.1(i)), obstructing inspectors (Article 38.1(i)), falsifying documents (Article 38.1(o)), product labeling (Article 27.1(b)), and capacity plans (Article 25.10(b)) during a port inspection in Randle-Galicia, Spain, on July 16, 2013.

Spain provided updates on each vessel in question. The *Puente Sabaris* was found guilty for the events of March 2013 and received a 12,100€ fine. The more recent allegations of May 2014 resulted in a three-month suspension from NAFO waters while the proceedings to sanction are underway. The *Villa de Pitanxo* (presumably CallSign 8) was found guilty for the events of February 2013 and received a fine of 113,702€. The vessel was also found guilty of the events of July 2013 and received a 9,601€ fine. The *Patricia Sotelo* (presumably CallSign 6) was found guilty for the events of July 2013 and received a fine of 289,001€, 10 points, and the seizure of 92,914 kilograms of Greenland halibut.

Since Spain provided information indicating the allegations concerning the three vessels in question were investigated and sanctioned where appropriate, Spain is not being identified.

IV. CERTIFICATIONS UNDER SECTIONS 609 AND 610

NMFS identified 10 countries in the 2013 Report to Congress as having vessels engaged in IUU fishing activity. Each incident of IUU fishing involved a violation of the rules of an international fishery management organization in 2011 or 2012. NMFS also identified one of those nations for bycatch activities. Under Section 609 of the Moratorium Protection Act, the Secretary of Commerce must certify biennially in the report to Congress whether an identified nation has taken appropriate corrective action to address the activities for which it has been identified.

After notifying the 10 countries of their identifications early in 2013, the U.S. Government consulted extensively with those governments, through face-to-face meetings, teleconferences, and correspondence, through late 2014. The 10 governments provided information that falls into several categories:

- For each of the acknowledged violations, the nations took punitive action against the vessels or persons (captains or vessel owners) involved, or explained why such action was not taken. The sanctions included fines, revocation of licenses, and forfeiture of catch and gear. An example is Italy, which seized and destroyed illegal driftnets, imposed fines on vessel owners using such nets, and initiated proceedings to suspend fishing licenses.
- As applicable and pertinent, the nations produced documentation of laws and regulations designed to combat IUU fishing, including measures that had recently been enacted or amended to give the nations more authority over their fishing fleets. For example, the Republic of Korea amended its Distant Water Fisheries Development Act (DWFDA) to increase fines and penalties for IUU fishing, and to make IUU fishing a criminal offense.
- In a few instances, nations did not provide evidence disputing the violations by their vessels in time to prevent their identification in the 2013 Report to Congress, but during the post-identification

consultation period offered credible evidence and explanations, based on investigations, that the vessels had not actually violated international measures. One example is Mexico's investigation of vessels alleged to have illegally discarded tuna, which resulted in the conclusion that the tuna had been determined unfit for human consumption and thus was allowed to be discarded.

The remainder of this section sets out in detail the information supplied by the identified nations about corrective actions taken – including penalties, withdrawal of fishing authorizations, and new fisheries management laws adopted – and NMFS' certification decisions for each nation. This process, as in past cycles of identifications and certifications, continues to operate as Congress intended: it is promoting compliance with international fisheries measures.

A. Colombia

Bases for 2013 Identification

Colombia was identified for having a number of vessels that violated IATTC resolutions in 2011. Three Colombian vessels, *Cabo de Hornos*, *Maria Isabel C*, and *Nazca*, finned sharks and discarded the carcasses at sea before the point of first landing in 2011, in violation of Resolution C-05-03. The *Cabo de Hornos*, *Sea Gem*, and *Sandra C* discarded salt bags or plastic trash at sea in 2011, violating Resolution C-04-05 Rev 2.

Notification and Consultation

Colombia was notified through a diplomatic note from DOS, dated January 10, 2013, and a letter from Samuel Rauch, the Deputy Assistant Administrator for Regulatory Programs, performing the functions of the Assistant Administrator for Fisheries, also dated January 10, 2013, of its identification as a nation whose vessels engaged in IUU fishing activity. The Embassy of Colombia in Washington was the primary entity involved in the consultation. The following lists the key communications between Colombia and the United States during the two-year consultation:

- On January 23, 2013, officials from NMFS and DOS met with the Colombian Embassy to discuss the identification of Colombia and the consultation process.

- On December 26, 2013, Colombia's Ambassador to the United States, Dr. Luis Carlos Villegas, sent a letter to NMFS describing initial actions taken to address the activities for which Colombia was identified.
- On July 22, 2014, NOAA received a letter from Alfonso Cuellar, Deputy Chief of Mission at the Embassy of Colombia in Washington, explaining the actions that had been taken to date regarding the vessels for which Colombia had been identified.
- On December 10, 2014, Colombia sent a response to NMFS with documentation of sanctions issued for the shark finning cases and additional information on the cases associated with trash or plastic discards.

Vessel-Specific Actions

Colombia opened administrative investigations into the owners, license holders, and captains of the three vessels that engaged in shark finning. According to confirmed observer reports, the *Cabo de Hornos* finned and discarded 24 sharks; the *Maria Isabel C* and the *Nazca* each finned and discarded one shark.

On December 5, 2014, final administrative decisions were issued, sanctioning the vessels with fines for violating Colombian law that prohibits the finning and discarding of sharks. The owner, commercial fishing permit holder, and captain of the *Cabo de Hornos* were fined 147,840,000 Colombian pesos (approximately $60,500) for finning 24 sharks. The owner, commercial fishing permit holder, and captain of the *Maria Isabel C* were fined 6,160,000 Colombian pesos (approximately $2,500) for finning one shark. The owner, commercial fishing permit holder, and captain of the *Nazca* were fined 6,160,000 Colombian pesos (approximately $2,500) for finning one shark.

The General Maritime Directorate, the authority that addresses discards of salt bags or plastic trash at sea, has opened preliminary investigations into the *Cabo de Hornos*, *Sea Gem*, and *Sandra C.* The Harbormaster of the Port of Cartagena is responsible for the investigations since the vessels are registered there. These investigations will determine whether there is sufficient evidence to initiate sanctions.

In order to address marine pollution, Colombian authorities implement procedures to prevent marine pollution from vessels and educate captains and crews regarding regulations, including the prohibition on improperly discarding trash, in compliance with Annex V of the International Convention for the Prevention of Pollution from Ships (MARPOL) and the relevant IATTC measure.

Colombia has also launched a campaign, "I am a guardian of the sea and river," to promote compliance with environmental protection and safety regulations. One element addresses the unlawful disposal of trash at sea. In addition, in 2013, the Ministry of Environment and Sustainable Development conducted three workshops along the Pacific coast of Colombia to address the problem of marine pollution.

Fisheries Management Measures

Resolution No. 375, adopted on April 17, 2013, requires that fins be naturally attached to ensure accurate identification of species at the first port of landing. Resolution No. 653 (September 7, 2012) allows Colombia to adopt management measures and make organizational changes to ensure the sustainability of fish stocks managed under the IATTC, and to amend national laws and regulations to ensure compliance with IATTC measures. Colombia has used this resolution as a basis for reviewing and revising applicable laws and regulations. It also adopted procedures to monitor and control compliance and has established rigorous penalties for violations of the General Fishing Statute.

Certification

Colombia sanctioned the owners, commercial fishing license holders, and captains of the *Cabo de Hornos*, *María Isabel C*, and *Nazca* for shark finning. Regarding the cases of the *Cabo de Hornos*, *Sea Gem*, and *Sandra C,* Colombia has opened preliminary investigations into the discards of trash, and conducts outreach efforts to fishermen and recreational boaters to educate them about the requirement for compliance with MARPOL Annex V. NMFS has determined that the Government of Colombia has taken appropriate corrective action to address the IUU fishing activities for which it was identified in the 2013 Report to Congress. Based on this finding, NMFS has made a positive certification determination for Colombia.

NMFS has again identified Colombia in this report (Part III.A.1) due to new IUU fishing activities conducted by the Colombian fleet during 2013 and 2014.

B. Ecuador

Bases for 2013 Identification

Ecuador was identified because a number of Ecuadorian-flagged vessels violated IATTC resolutions in 2011 and 2012. The *Drennec* finned 14 sharks and discarded the carcasses in violation of Resolution C-05-03. The following 11 vessels discarded salt bags or plastic trash at sea in 2011 in violation of Resolution C-04-05 Rev 2: the *Drennec*, *Lucia T*, *Rodolfo X*, *Zalbidea J*, *Monteneme*, *Yolanda L*, *Don Mario*, *Carmen D*, *Rosa F*, *Yelisava*, and *Ugavi Dos*.

On November 28, 2011, the fishing vessel *North Queen* traveled from Manta to Guayaquil without communicating a transit waiver to the IATTC Director, so the IATTC does not know whether Ecuador granted the waiver as required by Resolution C-09-04. The following seven vessels had interactions with sea turtles in 2011 without fully complying with the provisions of Resolution C-04-05 Rev 2, in that they failed to release the turtles: the *Gloria A*, *Via Simoun*, *Lucia T*, *Malula*, *Esmeralda C*, *Julia D*, and *Guayantuna I*. In January 2012, the *Julia D* made 13 sets less than a mile from a data buoy, in violation of Resolution C-11-03.

The following 16 vessels discarded tuna in 2011 in violation of Resolution C-11-01: the *Rocio*, *Charo*, *Rosa F*, *Julia D*, *Medjugorje*, *San Andres*, *Rossana L*, *Panchito L*, *Don Ramon*, *Via Simoun*, *Cap. Berny B*, *Pacific Tuna*, *Dona Roge*, *Esmeralda C*, *Sansun Ranger*, and *Ciudad de Portoviego*. A total of 57 sets had discards amounting to a total of 216.1 tons of tuna.

Notification and Consultation

Ecuador was notified through a diplomatic note from DOS, dated January 10, 2013, and a letter from Samuel Rauch of the same date, regarding its identification as a nation whose vessels engaged in IUU fishing activity. The Ministry of Aquaculture and Fisheries was the primary entity in Ecuador involved in the consultation. The following lists the key communications between Ecuador and the United States during the two-year consultation:

- On April 12, 2013, officials from NOAA and DOS met with officials from the Ministry of Fisheries via video-conference to discuss the identification of Ecuador and the consultation process.
- In June 2013, U.S. officials met with Guillermo Morán, Vice Minister of Aquaculture and Fisheries, to discuss the reasons Ecuador was identified and Ecuador's process in resolving cases.

- On September 25, 2013, Rodney McInnis, Acting Director of NMFS' Office of International Affairs, sent a letter to Mr. Morán requesting that, for those cases not under appeal, Ecuador send investigation or sanction updates, as applicable.
- On January 22, 2014, Mr. Morán sent a letter to Mr. McInnis explaining the actions taken against the vessels for which Ecuador was identified and the new Ministerial Agreement Ecuador has put into place. Ecuador also sent a copy of its report, "Review of the regulations and resolutions of IATTC and AIDCP [Agreement on the International Dolphin Conservation Program] governing tuna fishing."
- On January 29, 2014, Ecuadorian officials provided NOAA personnel with a letter explaining actions taken and the materials given to Ecuadorian vessel captains to ensure their awareness of IATTC resolutions at a workshop held in the summer of 2013.
- On July 17, 2014, Ecuadorian officials gave U.S. representatives a new Ministerial Agreement that implements IATTC Resolution C-13-01 as national law, along with details regarding shark and transiting regulations.
- On October 25, 2014, Ecuador provided more information on several of the cases, along with a copy of its shark regulations.

Vessel-Specific Actions

Ecuador opened administrative proceedings to sanction all the vessels charged with IATTC management measure violations, and reported that those charged presented two defenses: expiration of the 60-day statute of limitations, and the absence of any violations of national law because the applicable IATTC measures had not been implemented in Ecuador. The Director of Control of Fishery Resources found the defenses asserted to be legally sound and dismissed the administrative proceedings. See Part III.A.2 for an explanation of the statute of limitations issue.

Fisheries Management Measures

Ecuador has amended its regulations to implement all IATTC resolutions, including by providing penalties for violation of IATTC measures. In addition, Ecuador has informed both the IATTC and NOAA that it is working on new fisheries legislation, which presumably would address the short statute of limitations currently in place.

Additional Information

In the IATTC, the review panel that considers observer information to determine whether a violation has occurred looks at information from the previous year. Given Ecuador's statute of limitations of 60 days, this creates a timing problem. In an attempt to remedy the situation, Ecuador has entered into an agreement with the IATTC Secretariat under which it will be provided copies of all observer reports that identify violations before the reports pass through the review panel. The United States has expressed concern to Ecuador that a 60- day statute of limitations impedes meaningful fisheries enforcement and has urged the Government of Ecuador to significantly expand this time frame.

Ecuador conducted education and outreach during the closed fishing seasons in 2013 and 2014 to ensure that vessel captains are aware of IATTC resolutions and the repercussions of noncompliance. The training binders, with comprehensive materials, were sent to NOAA. They include an overview of the state of the world's fisheries, a synthesis of the IATTC and the AIDCP, statistics of tuna fisheries and catch, pertinent resolutions from the IATTC and potential violations by Ecuadorian vessels, tuna tracking information, turtle recovery and dis-entanglement methods, and shark conservation measures in place in Ecuador.

Certification

As noted above, Ecuador took action to sanction vessels charged with the IATTC violations that were the bases of its 2103 identification. The proceedings were dismissed, however, due to successful legal defenses. It is NMFS' understanding that this was the first time the statute of limitations has been used to challenge Ecuadorian actions for fisheries violations. In response, Ecuador acted to address the situation in the short term by amending its regulations to implement IATTC resolutions; NMFS understands that all IATTC resolutions are now in force in Ecuador. Ecuador also entered into an agreement with the IATTC Secretariat to expedite access to observer reports that identify possible violations. See Part III.A.2 for further information. In addition, Ecuador is currently working to draft new fisheries legislation that presumably would address the statute of limitations.

Taking into consideration the successful legal defenses to Ecuador's proceedings, NMFS has determined that the Government of Ecuador has taken appropriate corrective action to address the IUU fishing activities for which it was identified in the 2013 Report to Congress. Ecuador ensured that all IATTC measures have been implemented and took action to expedite access to

IATTC observer reports in an effort to identify violations within the 60-day statute of limitations. Based on this finding, NMFS has made a positive certification determination for Ecuador.

NMFS, however, has identified Ecuador in this report for IUU fishing activities conducted by the Ecuadorian fleet during 2013 and 2014 (Part III.A.2). While Ecuador implemented new IATTC regulations, they could not be applied retroactively to the 2013 and 2014 fishing activities for which it was identified. NMFS is concerned that Ecuador will not be able to sufficiently implement and enforce internationally agreed CMMs due to its short statute of limitations.

C. Ghana

Bases for 2013 Identification

Ghana was identified for failing to manage its fishing vessels consistent with measures adopted by ICCAT. The United States was specifically concerned about the following: data reporting and fleet control deficiencies, including data not submitted and data submitted late (Recommendation 05-09); overharvest of species, specifically the record of extensive overharvest of bigeye tuna (Recommendation 11-01); non-compliance with fleet capacity provisions (Recommendations 04-01 and 11-01); and Ghana's failure to implement effective measures to prohibit at-sea transshipments (Recommendation 06-11). With regard to capacity, Ghana needed to phase out two more purse seiners or four bait boats to meet ICCAT capacity limitation requirements in the bigeye tuna fishery.

While Ghana had prohibited at-sea transshipments, NMFS needed further evidence of the implementation and effective enforcement of these regulations. In addition, information indicated that Ghana had been overharvesting bigeye tuna since quotas were first imposed in 2004. ICCAT first identified Ghana for overharvest in 2009. NMFS believed that Ghana needed to show progress in complying with ICCAT recommendations by implementing the agreed payback plan for the overharvest of bigeye tuna, improving data collection, and complying with capacity limits. Improvement in the accuracy of Ghana's catch estimates was also needed to improve the ICCAT assessment of bigeye tuna stocks.

Notification and Consultation

Ghana was notified through a diplomatic note from DOS, dated January 10, 2013, and a letter from Samuel Rauch of the same date, regarding its identification as a nation whose vessels engaged in IUU fishing activity. The Ministry of Fisheries and Aquaculture Development was the primary entity within Ghana involved in the consultation. The following lists the key communications between Ghana and the United States during the two-year consultation:

- In November 2013, NOAA held two meetings with Ghana about improvements in Ghana's fisheries management.
- In February and March 2014, Ghana sent deletion certificates for a number of vessels and a copy of its Ministerial Directive prohibiting transshipment.
- On May 14, 2014, Ghana sent additional details and information on actions taken to address the activities for which Ghana was identified, including actions taken on bigeye tuna, swordfish overharvest, and fleet capacity control.
- On June 17, 2014, officials from NOAA and DOS met with Ghanaian officials to discuss Ghana's VMS and development of its fisheries management plan.
- On September 3, 2014, Ghana sent additional detail on its fleet capacity.
- On November 7, 2014, Ghana sent documentation outlining its compliance with capacity measures in ICCAT, and its progress on a fisheries management plan to address fishing capacity.

Data Reporting

Ghana submitted its compliance tables to ICCAT in 2013 on time; at the 2013 annual meeting, the SCRS commended Ghana for improvements in data reporting. In its annual report to ICCAT, Ghana stated that it continues to evaluate statistics based on improved sampling and provision of logbook and observer data.

Ghana currently collects data from vessels upon arrival in Ghanaian ports and at local canneries to improve catch composition reporting. Ghana's observer coverage has included six vessels for several months each year since 2009; in 2013, four purse seine vessels were covered year-round. All purse seine vessels had national observer coverage during the moratorium on fish aggregating devices in January and February of 2014.

The sampling bias detected in July 2012 has been addressed, so that Ghanaian sampling is now consistent with ICCAT sampling in terms of size structure and species composition. In addition, random sampling of landings has been implemented, and undocumented landings in Abidjan have been controlled through improved VMS and observer coverage. Ghana adopted the EC sampling protocol for logbook sampling, consistent with the ICCAT recommendation, with coverage now over 90 percent. Ghana is also taking steps to hire more samplers, to improve total coverage on vessels and to have the requisite number of trained technicians. Specifically, Ghana will have two samplers on each of its vessels that fish and land catch outside of Tema.

Fleet Control Deficiencies

Ghana established a fisheries enforcement unit, made up of the Navy, Air Force, Fisheries Commission, Attorney General's Department, Marine Police Unit, and the Bureau of National Investigations. It conducts operations 24 hours a day at sea and also has land-based and inland water patrols. Ghana's Maritime Authority originally had two employees, but now there are about 30.

Overharvest of Species

Ghana's payback plan for bigeye tuna, to compensate for past overharvests, was approved at the 2012 ICCAT meeting. Ghana was required to pay back 337 metric tons annually from 2012 through 2021. Transfers of 230 metric tons of bigeye from China, Chinese Taipei, Japan, and Korea went to Ghana to cover the 2012-2015 period. In 2011 and 2012, Ghana reported that it fished less than its annual quota and will continue to reduce its efforts to ease pressure on bigeye stocks. Specifically, at the 2013 annual meeting, the United States learned that Ghana had abided by its payback plan in 2012.

While the Letter of Concern issued by ICCAT following the 2013 meeting highlighted continued overharvest of South Atlantic swordfish, Ghana has shown a continuing trend of decreasing catches of this species. Ghana is implementing strategies to manage any overharvest of swordfish, such as giving fishermen gear to harvest semi-pelagic species in the inshore exclusive zone, and intends to work with ICCAT to review its swordfish allocation, since this artisanal fishery cannot be managed in the same manner as an industrialized fleet. Ghana also intends to request the transfer of underused swordfish quota from ICCAT CPCs.

Non-compliance with Fleet Capacity Provisions

According to Letters of Concern sent by ICCAT to Ghana after the 2011 and 2012 annual meetings, Ghana is required to reduce its purse seine fleet from 17 to 13 vessels. This capacity limit applies to vessels 20 meters length overall or greater fishing in the Atlantic bigeye fishery. Ghana is allowed to change the number of its vessels by gear type within its capacity limits communicated to ICCAT in 2005, on the basis of two bait boats for one purse seiner, with changes being approved by the Commission.

In its 2005 annual report to ICCAT, Ghana indicated that its bigeye tuna fleet was composed of 10 purse seine vessels and 26 bait boats. During the consultation, the United States learned that Ghana had 17 large-scale purse seine vessels in its fleet and 20 bait boats. During the consultation period with Ghana, the United States calculated that Ghana needed to remove only two bait boats or one purse seiner from the Atlantic bigeye tuna fishery in order to meet the requirements of Recommendation 11-01 based on Ghana's 2005 capacity data and the number of Ghanaian-flagged vessels on the ICCAT authorized vessels list. In communications with the United States, Ghana confirmed this was also its understanding. The *Delali*, an ICCATregistered purse seine vessel, sunk on December 24, 2013. Ghana has stated the vessel will not be replaced within its registry, allowing Ghana to be in compliance with its capacity limit.

Ghana has a general moratorium on flagging new vessels, is conducting an audit of industrial and semi-industrial vessels in its fleet, has set a goal to de-list inactive vessels and revoke the fishing licenses of those not in compliance, and is registering all canoes in order to gain a rough estimate of their fishing effort. This work is expected to be completed in mid-2015.

New fishing licenses, including replacement of old vessels, have been suspended in Ghana since February 1, 2012. A fishing vessel registry and licensing system database has been developed, and includes the following components for each vessel: identification, license information and dates of validity, MCS inspection reports, and vessel compliance reports and actions taken.

Transshipment

The two Ghanaian-flagged vessels that were at the center of the transshipment issues now have observers onboard. The vessels' movements have been restricted to port-to-port operations only, and they cannot deliver supplies to tuna or other fishing vessels at sea. Failure to comply with these provisions will result in cancellation of vessel licenses and fines. All Ghanaian large-scale tuna fishing vessels (over 24 meters) can transship only at authorized ports, under supervision. According to Ghana, there is full compliance with these provisions; enhanced VMS monitoring has provided an additional tool to ensure compliance with the prohibition.

Additional Information

Ghana is working to monitor tuna off-loadings, develop methods to measure the volume of off-loadings, and will require the cooperation of tuna fishing vessels as a requirement of their license. Working through the GEF and the FAO project on Areas Beyond National Jurisdiction, Ghana will also focus on MCS measures. The Ghanaian portion of this multi-national project will focus on installment of electronic monitoring systems and training on purse seine vessels.

Ghana has stated that bigeye tuna management is a challenge due to the association of bigeye with skipjack and yellowfin tuna catches. Because of this, Ghana is going to participate in research to mitigate the catch of bigeye tuna while fishing for skipjack. Ghana is also working to submit a management plan for fish aggregating devices to ICCAT by mid-2015, and intends to begin collecting logbooks on use of these devices earlier than the January 1, 2015, requirement.

In the medium term, Ghana is exploring two options to manage its bigeye tuna quota. The first is to more closely monitor its bigeye fishery and to close it when or just before the total allowable catch is reached. The second option is to give individual transferable quotas to licensed tuna vessels, which will equal the national total allowable catch.

Ghana has stated that additional adjustment to vessel capacity will be accomplished through the fisheries management reform that is being undertaken through the World Bank- and GEF-funded West Africa Regional Fisheries Program, due to be completed in 2017. As part of that project, Ghana is working on its fisheries management plan, which will include a fleet management strategy and promotion of alternative livelihoods such as aquaculture; it is expected to be completed by the end of 2015. Other efforts under that project include stock assessments to ensure appropriate management decisions, and pending legislation that will combat IUU fishing and include high monetary sanctions and the ability to suspend or revoke fishing licenses. Ghana recently collaborated with the FAO to obtain a contract with a Norwegian research vessel to conduct a stock assessment of Ghana's fisheries in September and October 2014; this work will assist Ghana in addressing overcapacity in its fisheries.

Certification

On the basis of the information provided on data reporting, control of the fishing fleet and at-sea transshipments, capacity reduction, and overharvest of species, NMFS has determined that the Government of Ghana has taken appropriate corrective action to address the IUU fishing activities for which it was identified in the 2013 Report to Congress. Therefore, NMFS has made a positive certification determination for Ghana.

D. Italy

Bases for 2013 Identification

In 2009, Italy was identified for several different violations of ICCAT requirements, including driftnet use, and in 2011 for additional driftnet fishing violations. The United States determined that Italy took appropriate corrective action to address each of the violations for which it was identified in 2009 and 2011, so it received positive certifications in the 2011 and 2013 Reports to Congress. Italy was then re-identified in the 2013 Report for ICCAT driftnet violations by different vessels in violation of ICCAT Recommendation 03-04.

While Italy had made progress in reducing illegal driftnet practices through enforcement actions against individual vessels and adoption of new laws to better address these activities, at least 18 new infractions were observed during the relevant time period of 2011–2012 for the 2013 identification. Several European Commission (EC) inspections in Italy in 2011 noted driftnet infractions of EC regulations and ICCAT measures. The inspectors found illegal nets on docks and listed a number of vessels either with driftnets or with gear typical of driftnet fishing, including nets longer than permitted and mesh size larger than allowed. EC inspectors also concluded that logbooks suggested underreporting and the capture of prohibited species while using driftnets. This information indicated that some vessels were still using illegal driftnets.

Italy subsequently investigated and sanctioned the vessels with driftnet violations in 2011, including by suspension of fishing licenses. Concerns remained, however, over the use of driftnets by Italian-flagged vessels. Given that illegal driftnet use by Italian-flagged vessels has been a long-standing issue, and that driftnet violations were again observed during the relevant time period, NMFS identified Italy in the 2013 Report to encourage Italy to end illegal driftnet use and to continue monitoring, surveillance, and control of its fishing fleet.

Notification and Consultation

Italy was notified through a diplomatic note from DOS, dated January 10, 2013, and a letter of the same date from Samuel Rauch, regarding its identification as a nation whose vessels engaged in IUU fishing activity. The Maritime Fisheries and Aquaculture Division under the Ministry of Agriculture, Food and Forestry was the primary entity within Italy involved in the consultation. The following lists the key communications between Italy and the United States during the two-year consultation:

- On January 16, 2013, an official from U.S. Embassy Rome met with Italian officials to initiate consultations under the Moratorium Protection Act.
- On September 18, 2013, Italian officials provided an overview of their continuing action against IUU fishing in a meeting with the Department of Commerce's Fisheries Representative to the EU, and officials from U.S. Embassy Rome.
- Italy sent regular enforcement reports throughout the consultation, detailing driftnet violations and sanctions by the Government of Italy.
- On July 15, 2014, an official from U.S. Embassy Rome met with Dr. Riccardo Rigillo, Director General of Fisheries in the Ministry of Agriculture, to learn about additional actions Italy has taken in eradicating illegal driftnets.

Vessel-Specific Actions

The Italian Port Captaincy and Coast Guard provided both enforcement data and sanctioning information to the United States on a regular basis. When illegal driftnets are found on Italian fishing vessels, the nets are seized by the Italian Coast Guard and destroyed, vessel owners are fined, and fishing licenses are suspended. Legal nets used in an illegal manner are also confiscated; vessel owners can petition for their return after fines have been paid.

Eighteen violations with *ferrettara* (small-mesh driftnets) occurred in 2013, all for possessing driftnets that were longer than the 2.5 kilometers allowed under regulations. Five of the violations were for nets found abandoned on docks where the owner could not be identified; these nets were confiscated. The remaining 13 offenders were fined, their nets seized, and their fishing licenses suspended for three months, with the exception of the *Madonna di Fatima*, which received a longer suspension due to a prior driftnet violation. Fines ranged from 4,000€ to 10,000€.

Thirteen *ferrettara* violations occurred in 2014. Five of the cases dealt with illegal nets found at sea without an identifiable owner; each net was seized. One case involved discovery on a truck of a net of illegal length. Three cases involved illegal nets with mesh size larger than permitted by regulations. The four remaining cases dealt with illegal nets of longer than allowed length and mesh size greater than permitted under regulations. Each of the eight owners was fined 4,000€ and the nets seized. Fishing license suspensions are pending.

Additional Information

During the September 18, 2013, meeting with U.S. Embassy Rome staff, Commander Vittorio Giovannone explained that Italy is drafting new provisions that would increase the fines for illegal fishing, in response to concerns about overfishing of specific species. The Coast Guard is also considering action to further reduce the number of ships using long lines and *ferrettara* nets.

The Coast Guard is also determining how to track small boats (shorter than 12 meters), which are not required to have a VMS on board or to submit e-logbooks. A new Coast Guard ship, specifically designed for control activities and scientific research, was under construction in Naples and will soon be ready to commission for service.

In 2011, the EC called on Italy to comply with the European Court of Justice ruling of October 2009 to end the illegal use of driftnets. The EC legal and policy review of Italy's response looked into the level of sanctions applied to IUU fishing and legislative initiatives against driftnets, among other things. A Driftnet Action Plan, as required by the EC, was designed to overcome the shortcomings in the Italian fisheries control system. The Action Plan, issued in December 2013, had been 80 percent implemented as of July 2014. It includes requirements for more detailed monitoring of activities at the local level and updates to national and local enforcement plans. Because Italy has addressed the driftnet issue sufficiently, as evidenced above, the EC has closed its driftnet infringement proceedings against Italy.

Certification

On the basis of information provided on sanctions against illegal driftnet use and other actions taken to end this practice, NMFS has determined that the Government of Italy has taken appropriate corrective action to address the IUU fishing activities for which it was identified in the 2013 Report to

Congress. Based on this finding, NMFS has made a positive certification determination for Italy.

E. Korea (Republic of)

Bases for 2013 Identification

The Republic of Korea was identified for failing to apply sufficient sanctions to deter its vessels from engaging in fishing activities that violate binding measures adopted by CCAMLR. In 2011, a Korean fishing vessel (the *Insung No. 7*) set fishing gear in CCAMLR waters, even though the master knew that the catch limit had already been exceeded and the area had been closed to fishing. Illegal fishing by the *Insung No. 7* in Division 58.4.2 resulted in the area's catch limit being exceeded by 339 percent, with the illegal catch estimated by Korea to be worth $710,000. Korea imposed a fine of approximately $1,300 and a 30-day suspension of the vessel's distant water fishing authorization. Korea also reported that the vessel master's license might be suspended for 30 days.

At the 2011 CCAMLR meeting, however, many delegations, including the United States, viewed Korea's sanctions against the operator, vessel, and master as inadequate, given the seriousness of the illegal activity. CCAMLR's Standing Committee on Implementation and Compliance proposed placing the vessel on the Contracting Party IUU Vessel List, but Korea blocked its inclusion. Although Korea agreed at the 2011 CCAMLR meeting to withdraw three of its vessels from the CCAMLR toothfish fishery for the 2011–2012 fishing season as a concession, this decision did not rectify the inadequacy of Korea's enforcement measures to deter future violations. At the time, Korea indicated that it was undertaking amendment of its law to strengthen the sanctions available to address IUU fishing activities. In reviewing the situation as part of the Moratorium Protection Act identification process, the United States maintained the view that, if Korea's pending amendment took effect, it would most likely still be insufficient to deter Korean vessels from violating measures adopted under an international fisheries management agreement. In particular, the pending amendment did not appear to give Korea the ability to seize illegal catch or its proceeds.

Given Korea's inability to effectively control its fishing vessels as demonstrated, in part, by the minimal sanctions Korea placed on vessels found to be carrying out IUU fishing, NMFS identified Korea in the January 2013 Report to Congress.

Notification and Consultation

The Republic of Korea was notified through a diplomatic note from DOS, dated January 10, 2013, and a letter of the same date from Samuel Rauch, regarding its identification as a nation whose vessels engaged in IUU fishing activity. The Ministry of Foreign Affairs and the Ministry of Oceans and Fisheries were the primary entities involved in the consultation. The following lists the key communications between Korea and the United States during the two-year consultation:

- On May 31, 2013, the Ministry of Oceans and Fisheries sent correspondence to NMFS outlining tools it can use in the fight against IUU fishing and its higher penalty schemes that would be included in an amendment to its DWFDA.
- In July 2013, on the margins of the ICCAT Convention Amendment Working Group in Sapporo, Japan, NOAA's Deputy Assistant Secretary for International Fisheries, Russell Smith, met with Jeongseok Park, Fisheries Negotiator in the Ministry of Oceans and Fisheries, to discuss IUU fishing and Korea's draft amendments to its DWFDA.
- On September 3, 2013, NMFS received a copy of Korea's final DWFDA from U.S. Embassy Seoul.
- On February 27, 2014, Korea sent several documents to Mr. Smith, including the DWFDA Enforcement Rules and the Ministerial Directive on Port State Inspections.
- On April 7, 2014, the Minister for Economic Affairs, Gheewhan Kim, met with Mark Schaefer, Assistant Secretary of Commerce for Conservation and Management, to discuss Korea's DWFDA and its implementation.
- On April 22, 2014, DOS officials met with officials from the Korean Embassy in Washington; Korea outlined measures it had taken to address IUU fishing, and agreed to share information on any new cases of IUU fishing that would be addressed under the newly implemented DWFDA.
- On May 5, 2014, Korea sent a spreadsheet showing sanctions that had been imposed on 22 Korean-flagged vessels that committed IUU fishing in West Africa between July 2011 and January 2014.
- On May 14, 2014, Mr. Smith met with officials from the Embassy of Korea to hear about efforts Korea is taking to end IUU fishing in West

Africa, including establishing a hotline for reporting and improving the review of vessel histories prior to issuing catch certificates.

- On July 18, 2014, Korea sent a draft of further amendments to the DWFDA and associated documents for the United States' review and comment.
- On September 19, 2014, a NMFS official met with representatives from the Embassy of Korea to review recent actions Korea has taken with regard to the *Insung* vessels and seven vessels said to have been fishing in Sierra Leone's waters.
- In November 2014, Korea provided a copy of its National Plan of Action to Prevent, Deter and Eliminate Illegal, Unreported and Unregulated Fishing, its port State inspection scheme, and catch certification suspension records.

Fisheries Management Measures

Korea passed the most recent amendments to the DWFDA on July 30, 2013, with an effective date of January 31, 2014. Throughout the process of developing its new amendments, Korea welcomed U.S. suggestions on how to proceed so that it could develop a robust anti-IUU fishing regime.

The new amendments make IUU fishing a criminal offense, in addition to being an administrative offense, and increase fines and penalties for IUU fishing. Violations include fishing without authorization, in a closed area or for prohibited species, without a quota or over quota, with prohibited or unauthorized gear, or in contravention of RFMO measures; transshipping with an IUU vessel; concealing or modifying ship markings; and capturing fish smaller than permitted size. Violations of these measures are subject to imprisonment of up to three years or criminal fines up to three times the value of the fisheries products. Catches, products, vessels, fishing gear, and other items may also be confiscated for these violations. Ship officers' licenses can be suspended or cancelled upon violations of the DWFDA. Suspension of fishing operations of up to six months can also occur, *inter alia*, when fishing permission is obtained fraudulently or under other circumstances, such as exceeding quotas, and for the specific violations mentioned previously.

Korea checks the IUU fishing and VMS records prior to issuing a catch certificate to a vessel. Korea had 100 percent VMS coverage of its distant water fleet in March 2014. If the VMS is not operational, Korea will not issue a catch certificate and will initiate an investigation. As of July 2014, Korea has denied catch certificates to 41 vessels based on their associations with IUU fishing. Korea also opened a fisheries monitoring center in Busan in March

2014. This center monitors the movement of high seas vessels, maintains VMS records, and facilitates VMS verification used in issuing catch certificates.

Korea is currently working on further amendments to the DWFDA to strengthen certain provisions, including control of its nationals on foreign-flagged vessels, treatment of vessels that commit IUU fishing, and strengthening of some sanctions.

Vessel-Specific Actions

Korea has found cases of IUU fishing, in addition to the *Insung No. 7*, that occurred before implementation of the DWFDA on January 31, 2014. An investigation of fishing activities in 2009-2011 within the Convention Area by three other *Insung*-owned vessels concluded that all three vessels were engaged in IUU fishing. Korea reported that it had imposed much stronger sanctions against the vessels: administrative fines; license suspension for the masters of two of the vessels and license revocation for the third master; and a 10-year fishing ban on the rest of the *Insung* fleet so it they cannot fish for toothfish in the Convention Area during that time.

Korea also provided a document outlining the sanctions imposed on 22 Korean-flagged fishing vessels that engaged in illegal fishing, transshipments, and unauthorized fishing in West African waters, including off Liberia, Guinea, and Sierra Leone in 2013 and 2014, before the amended DWFDA came into effect. Administrative fines and suspension of the officers' licenses and fishing authorizations were imposed, including longer suspensions if previous IUU fishing had been committed by that same vessel. The sanctions were the maximum that could be imposed under the old regulations since the activities took place before the new DWFDA was implemented.

Since passage of the amendments to the DWFDA, the United States has requested examples of how Korea now handles cases of IUU fishing. NMFS recently learned of seven cases of fishing by Korean-flagged vessels in Sierra Leone in 2014, where questions lingered as to the legality of the fishing licenses issued by that nation. Korea ordered the vessels to stop fishing in Sierra Leone's Inshore Exclusion Zone, and blocked the fish from entering Korean and other markets by refusing to validate catch documentation for the fish. Korea has issued a policy that restricts Korean-flagged fishing vessels from waters of coastal States with fishing authorization management systems that may be insufficient.

Certification

Korea has instituted new sanctions through amendments to its DWFDA. Korea also ensured installation of VMS on all of its distant water fishing vessels and opened its Fisheries Monitoring Center in 2014. Korea has actively educated its distant water fishing industry on IUU fishing and the new amendments to the DWFDA that are now in place. Korea is also engaging with West African States to encourage better monitoring and control of its fleet operating in West African waters. Korea has also finalized its National Plan of Action for IUU fishing, is working on new amendments to the DWFDA, engaging with NGOs to address IUU fishing concerns, particularly in West Africa, and has begun engaging more with the global fisheries community to address and solve issues with its distant water fleet.

NMFS has determined that the Republic of Korea has taken appropriate corrective action to address its failure to apply sufficient sanctions to deter its vessels from engaging in fishing activities that violate binding conservation and management measures adopted by CCAMLR, for which it was identified in the 2013 Report to Congress. Based on this finding, NMFS has made a positive certification determination for Korea.

F. Mexico

Bases for 2013 Identification for IUU Fishing

Mexico was identified for IUU fishing based on the activities of several Mexican-flagged fishing vessels that violated IATTC resolutions in 2011 and 2012. One vessel, the *Atún VII,* finned sharks and discarded the carcasses at sea, in violation of Resolution C-05-03. Eight Mexican vessels discarded salt bags or plastic trash at sea, violating Resolution C-04-05 Rev 2: the *Atún VII*, *Azteca 5*, *Bonnie*, *Buenaventura I*, *Cartadedeces*, *Chac Mool*, *Maria Luisa*, and *Nair*. Five vessels (the *Arkos I Chiapas*, *Atún VI*, *Azteca 10*, *Azteca 2*, and *Maria Rosana*) violated sea turtle bycatch mitigation measures stipulated in Resolution C-04-05 Rev 2 by failing to release turtles. The *Nair II* and *El Dorado* discarded seven tons of tuna over four sets in violation of Resolution C-11-01.

Bases for 2013 Identification for Bycatch of PLMRs

In 2012, Mexican fishing vessels in a groundfish fishery in Baja California Sur incidentally caught North Pacific loggerhead sea turtles, a PLMR shared with the United States. In October 2012, the Mexican Fisheries Research

Institute published a report on bycatch reduction trials conducted earlier in 2012 in the gillnet fishery in Baja California Sur. During six days of trials, 88 loggerhead sea turtles were captured. The report concluded that local fleets likely have high bycatch rates. In July and August 2012, 438 North Pacific loggerhead sea turtles stranded, dead, along 43 kilometers of the shoreline of Playa San Lazaro, Baja California Sur, according to Mexican Wildlife Law Enforcement. Mexico did not provide evidence of management measures that are comparable in effectiveness to those of the United States to address this bycatch.

Notification and Consultation

On January 10, 2013, Mexico was notified of its identification in the 2013 Report to Congress for IUU fishing and bycatch of North Pacific loggerhead sea turtles through a diplomatic note from DOS and a letter from Samuel Rauch. The National Commission of Aquaculture and Fisheries was the primary entity within the Mexican Government involved in the consultation. The following lists communications between Mexico and the United States during the two-year consultation period:

- On February 1, 2013, NOAA received information from Mexico regarding efforts and regulations to protect all sea turtles nationally and efforts in the Gulf of Ulloa to address loggerhead sea turtle strandings.
- On March 4, 2013, the United States requested information from Mexico on actions taken to address the IUU fishing cases and regulations to address bycatch of North Pacific loggerhead sea turtles in the coastal gillnet fishery of Baja California Sur, as well as on non-regulatory actions to conserve North Pacific loggerhead sea turtles.
- On May 22, 2013, Mexico provided information on alternative gear research that had been carried out by the Government.
- On July 2, 2013, Mexico provided information on actions taken in regard to the vessels identified for IUU fishing and efforts it had conducted dealing with North Pacific loggerhead sea turtles.
- On January 6, 2014, NMFS requested information on Mexico's actions to investigate the vessels identified for IUU fishing, as well as actions to address the bycatch of North Pacific loggerhead sea turtles.
- On May 12, 2014, officials from Mexico and the United States met to discuss Mexico's identification under the Moratorium Protection Act.

Discussion focused on the type of information the United States requires.

- On July 25, 2014, Mexico and the United States met at their annual Fisheries Bilateral in Mazatlán, Sinaloa, Mexico. Mexico outlined actions taken to address the violations of IATTC measures and actions it is taking related to North Pacific loggerhead sea turtles.
- On August 28, 2014, Mexico provided information on actions taken with regard to its vessels that violated IATTC conservation measures and steps being taken to address and investigate bycatch of loggerhead sea turtles.
- On October 13, 2014, Mexico provided additional information outlining the basis of its bycatch management measures and the program it is working on for the Gulf of Ulloa.
- On October 16, 2014, Mexican and U.S. officials discussed remaining needs in relation to the IATTC violations, and the program Mexico is working on to address bycatch of loggerhead sea turtles in the Gulf of Ulloa.
- On November 5, 2014, Mexico provided additional information on actions taken for the remaining IATTC cases, and additional details on the bycatch program.
- On December 10, 2014, Mexico sent additional background information on the bycatch program it is working on.

Vessel-Specific Actions

An investigation of the *Atún VII* (now the *Adriana M*) determined that infractions of shark finning were committed. Sanctions equal to 2,220,570 pesos (approximately $165,000) will be imposed on the owner of the vessel, subject to appeal.

The *Atún VII* (now the *Adriana M*), *Azteca 5*, *Bonnie*, *Buenaventura I*, *Cartadedeces, Chac Mool*, *Maria Luisa*, and *Nair* cases were turned over to the Office of the Federal Attorney for Environmental Protection, which determined that the acts were not punishable under domestic law given the small volume of material. The companies were notified to stop disposing of refuse at sea, since it violated IATTC provisions, through written warnings on November 5, 2012.

The cases against the *Arkos I Chiapas*, *Atún VI*, *Azteca 10*, *Azteca 2*, and *Maria Rosana* were investigated by the Office of the Federal Attorney for Environmental Protection, which determined that the acts were not violations because the turtles were released unharmed. Mexico supplied a summary of the observer statements and reports, indicating that the turtles in each case were released unharmed.

The *Nair II* and *El Dorado* discarded tuna because it was not fit for human consumption, as stated in observer reports and captains' statements. This result is consistent with IATTC Resolution C-12-01, which requires all purse seine vessels to retain onboard and land all bigeye, skipjack, and yellowfin tuna caught, except fish considered unfit for human consumption for reasons other than size.

Certification for IUU Fishing

Mexico provided documentary information on the actions taken to address the cases of violations of IATTC conservation and management measures. It sanctioned the *Atún VII* for shark finning, investigated the discarding of trash at sea and issued a warning to those vessels, investigated the sea turtle release cases and found that the turtles were released unharmed, and determined that the tuna discards took place because the tuna were unfit for human consumption. On the basis of this information, NMFS has determined that the Government of Mexico has taken appropriate corrective action to address the IUU fishing activities for which it was identified in the 2013 Report to Congress. Based on this finding, NMFS has made a positive certification determination for Mexico for this portion of its identification.

NMFS, however, is identifying Mexico in this report for IUU fishing activities conducted by Mexican vessels during 2013 (see Part III.A.3).

Certification for Bycatch

To receive a positive certification under the Moratorium Protection Act, a nation identified for bycatch must have adopted a regulatory program to end or reduce bycatch that is comparable in effectiveness to U.S. measures, and have established a management plan to assist in the collection of species-specific data to support stock assessment and conservation efforts. The Government of Mexico is undertaking efforts to conserve sea turtles, and in the latter half of 2014 made significant progress to develop a regulatory program to address the bycatch of North Pacific loggerhead sea turtles and to better manage the gillnet fisheries in the Gulf of Ulloa.

On December 5, 2014, Mexico published a proposal to establish a refuge area within the Gulf of Ulloa for the conservation of loggerhead sea turtles under the General Wildlife Act. The proposal was available for public comment for 20 days and then underwent a 20-day regulatory review process. After the review process, Mexico will publish the final notification that the refuge is in force. After the refuge is adopted, Mexico has reported that it will undertake a process, to be finalized within 180 days, to develop relevant fisheries management measures that address bycatch of North Pacific loggerhead sea turtles within the refuge area. Mexico has stated that it plans to adopt this program by the opening of the gillnet fishery in May 2015. NMFS does not yet have details on the specific elements to be included in the final regulatory program.

Because the Government of Mexico has proposed establishment of a refuge area and made progress toward adopting a bycatch regulatory program, NMFS is delaying its certification determination until May 2015. When making that determination, NMFS will consider whether the program and associated regulations, as finalized, are comparable in effectiveness to relevant U.S. regulatory measures. This delay in certification does not prejudice NMFS' final certification decision.

G. Panama

Bases for 2013 Identification

Panama was identified because several Panamanian-flagged vessels violated IATTC resolutions in 2011 and/or 2012. The *Delia* finned a shark and discarded the carcass prior to the point of first landing, in violation of Resolution C-05-03. Three vessels (the *Delia*, *Connie Jean Two*, and *El Marquez*) discarded salt bags or plastic trash, in violation of Resolution C-04-05 Rev 2. The *Contadora I* and *Delia* discarded tuna in violation of Resolution C-11-01; a total of 14 sets had discards amounting to 22.8 tons of tuna.

Notification and Consultation

Panama was notified through a diplomatic note from DOS, dated January 10, 2013, and a letter from Samuel Rauch of the same date, regarding its identification as a nation whose vessels engaged in IUU fishing activity. The Aquatic Resources Authority of Panama was the primary entity in Panama involved in the consultation. The following lists the key communications between Panama and the United States during the two-year consultation:

- On January 10, 2013, NMFS contacted Panamanian officials about the identification; they responded that information would be forthcoming.
- On August 8, 2014, Panama sent a letter to NOAA outlining actions it has taken against three of the four vessels for which it was identified.

- On October 15, 2014, the United States sent a letter to Panama requesting documentation of sanctions placed on the vessels for which it was identified, along with additional information.
- On November 6, 2014, Panama responded with documentation of actions taken against some of the vessels for which it was identified.
- On December 10, 2014, Panama sent documentation related to one of the vessels, as requested by NMFS.
- On December 16, 2014, Panama sent the last piece of documentation requested by NMFS related to the actions Panama has taken on the vessels for which it was identified.

Vessel-Specific Actions

Charges were filed against the *Delia* on July 5, 2012, for discarding tuna. A penalty ruling was issued containing a written warning that, if the same actions are repeated by the vessel, maximum fines would be imposed. Because this vessel is currently flagged to Ecuador under a chartering arrangement, Panama alerted Ecuador of the issue so that Ecuadorian authorities may take any necessary action. The letter alerted Ecuador to the investigation into the *Delia* for discarding tuna, but reported that no investigation into the shark finning violation or the discards of trash or plastic at sea had been initiated because the vessel was reported to be flagged under Ecuadorian, rather than Panamanian, flag at the time of charging.

Panama issued a fine of 1,000 Balboa ($1,000) against the *Connie Jean Two* for discards of salt bags or plastic trash. Panama investigated the *El Marquez*, and issued a fine of 2,500 Balboas ($2,500) for violations of discarding salt bags or plastic trash at sea.

Panama found sufficient evidence to open administrative proceedings, on November 17, 2011, against the *Contadora I* for discarding tuna. Sanctions were imposed on the vessel on June 20, 2013, in the amount of 10,500 Balboas ($10,500) for violating management measures and national laws.

Fisheries Management Measures

Panama passed Executive Decree No. 160 of June 6, 2013, which set forth procedures to impose administrative sanctions for violations of the regulations on aquatic, coastal/marine, and fishery resources included in Law 44 of November 23, 2006. A vessel desiring to change owners or cancel its registration must now pay any pending fines, or present a bond of $1 million. Before this decree came into force, vessels fined for violations could cancel their Panamanian registration and thereby avoid paying their fines.

Executive Decree No. 161 of June 6, 2013, provides the mechanisms of inspection, monitoring, and control of nationally registered fishing and fishing support vessels that operate internationally. Executive Decree No. 162 of June 6, 2013, establishes and regulates fishing and fishing support licenses for vessels that fish internationally.

In addition, the Panamanian agencies have implemented inter-institutional cooperation and now exchange information on fishing vessels and national and international fishery inspections. The automation of catch certificates has been regulated; fishing license information can now be verified online.

Certification

On the basis of information provided about actions against the vessels that violated IATTC resolutions, NMFS has determined that the Government of Panama has taken appropriate corrective action to address the IUU fishing activities for which it was identified in the 2013 Report to Congress. Based on this finding, NMFS has made a positive certification determination for Panama for its IUU fishing identification.

H. Spain

Bases for 2013 Identification

Spain was identified because two Spanish-flagged vessels engaged in fishing activities that violated CMMs required under an international fishery management agreement. One vessel violated NAFO conservation and enforcement measures; the other, an IATTC conservation and management recommendation. At the time of identification, Spain was investigating both vessels.

On February 3, 2012, Canadian NAFO inspectors boarded the *Pescaberbes Dos* in the NAFO Regulatory Area and reported that approximately 134.7 tons of product was not labeled, as required by a NAFO measure (Chapter IV, Article 24.1), and that approximately 30.2 tons of Greenland halibut product was not marked as having been harvested in NAFO Divisions 3KLMNO, a separate requirement (Chapter IV, Article 24.1). Spanish inspectors confirmed the violation related to the labeling of boxes. Spain explained that initial infringement proceedings against the vessel's owner had begun.

On August 18, 2011, the *Albacora Uno* discarded a ton of skipjack tuna in violation of IATTC Resolution C-11-01, which requires vessels to retain all catch of skipjack tuna. Spain was seeking information to determine the accuracy of the report.

Notification and Consultation

Spain was notified through a diplomatic note from DOS, dated January 10, 2013, and a letter from Samuel Rauch of the same date, regarding its identification as a nation whose vessels engaged in IUU fishing activity. The Secretariat General for Fisheries was the primary entity within Spain involved in the consultation. The following lists the key communications between Spain and the United States during the two-year consultation:

- On January 16, 2013, DOS and NMFS officials met with the Spanish Embassy in Washington, D.C., about the identification of Spain and the consultation process.
- On September 2, 2013, a representative of Spain's Secretariat General for Fisheries informed a NMFS official about actions taken in the cases of the two vessels for which Spain was identified; NMFS requested additional information.
- On December 20, 2013, a Spanish official provided further documentation of the actions taken by Spain.
- On April 25, 2014, NMFS sent a follow-up letter to Spain requesting documentation of the sanction applied to the *Pescaberbes Dos* and suggesting that the EU and the United States work with other IATTC members at its next annual meeting to clarify the IATTC discard rules.
- On June 23, 2014, Spain responded with documentation of the sanctioning of the *Pescaberbes Dos* and agreed that tuna discard rules within the IATTC need to be clarified.

Vessel-Specific Actions

Spain supplied documentation that a fine of 4,500€ had been imposed on the captain and owner of the *Pescaberbes Dos* for mislabeling boxes. On April 7, 2014, the Director General dismissed the administrative appeal by the company.

Spain reported that the file on the *Albacora Uno* was closed by Spanish enforcement authorities because there were insufficient grounds for a sanction on the alleged discard of skipjack tuna. The Spanish enforcement services investigated the claims of the observer and the findings of the IATTC, along with other relevant information, in coming to this conclusion.

Certification

The Government of Spain took corrective action for the two vessels identified for IUU fishing in the 2013 Report to Congress. In the case of the *Pescaberbes Dos*, Spain imposed a 4,500€ sanction. Spain closed the investigation surrounding the *Albacora Uno*, concluding that there were insufficient grounds for a sanction. In addition, Spain agreed on the need to clarify the rules with regard to tuna discards in the IATTC and stated that it will work with the EU to address the issue within the IATTC. NMFS has determined that the Government of Spain has taken appropriate corrective action to address the IUU fishing activities for which it was identified in the 2013 Report to Congress. Based on this finding, NMFS has made a positive certification determination for Spain.

I. Tanzania

Bases for 2013 Identification

Tanzania was identified in the 2013 Report to Congress because four of its vessels undermined the effectiveness of CCAMLR conservation measures. All four vessels are currently listed on the CCAMLR NCP-IUU Vessel List and are not permitted to fish in the Convention Area. Placed on the IUU vessel list in 2008, the *Wutaishan Anhui 44*, now known as the *Nihewan*, was observed inside Division 58.4.1 on January 20, 29, and 30, 2012, apparently flagged by Tanzania (the Zanzibar Maritime Authority is the competent authority regarding vessel registration). During the January 20 sighting, this vessel contacted a vessel authorized to fish in the Convention Area about their fishing lines being entwined, thus indicating the *Wutaishan Anhui 44* was also fishing in the Convention Area. Australian authorities again observed this vessel on April 24, 2012, in the vicinity of Christmas Island, northwest of Australia, still flagged by Tanzania with the same external markings. French authorities then sighted this vessel hauling unmarked bottom-set gillnet in Statistical Subarea 58.6 on July 3, 2012. This time, the vessel displayed the name *Huiquan* and a different international radio call-sign, still claiming Tanzanian registry with a homeport of Zanzibar.

The *Shaanxi Henan 33*, now known as the *Chengdu*, was observed by Australian authorities on May 16, 2012, in the vicinity of Christmas Island, apparently flagged by Tanzania (Zanzibar). Australian authorities noted that, while this sighting was outside the Convention Area, this vessel has a long history of association with IUU fishing inside the Area, has been on the IUU vessel list since 2004, and continues to undermine CCAMLR measures.

On April 1, 2012, Australian authorities observed the *Huang He 22*, now known as the *Chang Bai* and on the IUU vessel list since 2003, in the vicinity of Christmas Island, displaying a Tanzanian-allocated call sign. French authorities sighted this vessel in Statistical Subarea 58.6 on July 1, 2012. At the time of the second sighting, the vessel was underway with fishing gear visible on the deck and displaying the same external markings as before. Australian authorities noted that this vessel had been sighted on four occasions since 2004 and was suspected of violating CCAMLR conservation measures. This vessel had been listed on the IUU vessel list under nine other names and had also been reported as having been flagged to six other nations in the past.

The refrigerated cargo vessel *Baiyangdian*, now known as the *Tiantai*, was observed inside Division 58.4.1 on January 28, 2012. Inconsistencies between information provided by the captain and other sources, as well as the captain's lack of willingness to communicate with Australian authorities, raised concerns that the vessel may have been engaged in IUU fishing by supporting vessels suspected of IUU fishing activities in the Convention Area. These concerns were reinforced when this vessel was observed on April 1, 2012, under tow by the CCAMLR - listed IUU vessel *Huang He 22* in the vicinity of Christmas Island. As a result, the vessel was added to the CCAMLR NCP-IUU Vessel List.

Notification and Consultation

Tanzania was notified through a diplomatic note from DOS, dated January 10, 2013, and a letter from Samuel Rauch of the same date, of its identification as a nation whose vessels engaged in IUU fishing activity. NMFS worked through the U.S. Embassy in Dar es Salaam to facilitate consultation with Tanzania and Zanzibar. The following lists the key communications between Tanzania and the United States during the consultation:

- On January 24, 2013, the United Republic of Tanzania Ministry of Livestock and Fisheries Development sent a letter addressed to NMFS explaining the status of three of the four vessels for which it was identified.
- On March 15, 2013, the United States sent a follow-up letter to Tanzania requesting documentation of the three vessels and inquiring about the fourth vessel for which Tanzania had been identified.

- On March 3, 2014, the Zanzibar Maritime Authority supplied the U.S. Embassy in Dar es Salaam with documentation showing the four vessels are no longer registered to Tanzania.
- On April 24, 2014, NMFS informed the Zanzibar Maritime Authority those same four vessels had changed their names and were still using radio call signs associated with Tanzania. If these vessels have not reflagged, they are not entitled to use radio call signs associated with any flag State.

Vessel-Specific Actions

The Zanzibar Maritime Authority provided copies of the Provisional Certificates of Registry for the *Wutaishan Anhui 44*, *Shaanxi Henan 33*, and *Huang He 22*, indicating that the vessels are no longer entitled to fly the Tanzanian flag. The Tanzania Zanzibar International Register of Shipping sent a letter to the owners of the *Baiyangdian* removing the vessel from the registry. None of the vessels has been re-registered following the expiration of their provisional registrations.

Tanzania has stated that other Tanzanian-flagged vessels found to be involved in IUU fishing activities will also face de-registration. In addition, a vessel is not to be registered until the Tanzanian Deep Sea Fishing Authority confirms that it has no IUU fishing history.

Certification

Tanzania ensured the removal of the offending vessels from its registry, and is requiring future vessel registrations to be cross-checked for IUU fishing histories. In addition, Tanzania is exploring a contract with a new company to oversee the Zanzibar International Shipping Registry. NMFS has determined that the Government of Tanzania has taken appropriate corrective action to address the IUU fishing activities for which it was identified in the 2013 Report to Congress. Based on this finding, NMFS has made a positive certification determination for Tanzania.

J. Venezuela

Bases for 2013 Identification

Records from the IATTC indicate a number of Venezuela's fishing vessels violated IATTC resolutions in 2011. These violations include two vessels (the *Ventuari* and *Cayude*) that finned sharks and discarded the carcasses at sea, in violation of Resolution C-05-03. The *Don Francesco, La Rosa Mistica,* and *Taurus I* discarded salt bags or plastic trash at sea, in violation of Resolution C-04-05 Rev 2. The *Don Francesco* and *Curimagua* violated sea turtle bycatch mitigation measures by failing to release turtles, in violation of Resolution C-04-05 Rev 2. The *La Rosa Mistica*, *Amazonas,* and *Canaima* illegally discarded tuna, in violation of Resolution C-11-01, in a total of six sets amounting to 25 tons of tuna.

Notification and Consultation

Venezuela was notified through a diplomatic note from DOS, dated January 10, 2013, and a letter from Samuel Rauch of the same date, regarding its identification as a nation whose vessels engaged in IUU fishing activity. NMFS worked through the U.S. Embassy in Caracas to facilitate consultation with Venezuela.

- Venezuela informed the United States that it provided information regarding actions taken to address the IUU fishing activities for which it was identified to the IATTC Secretariat.
- On August 15, 2014, the United States requested information from the IATTC Secretariat on the 2011 vessel infractions for which Venezuela was identified in the 2013 Report to Congress.
- On September 18, 2014, the United States received from the IATTC Secretariat the information that Venezuela had submitted about its actions to address violations of IATTC measures that took place in 2011, as outlined below.

Fisheries Management Measures

The Socialist Institute of Fisheries and Aquaculture took measures to improve overall management of the Venezuelan fishing fleet. All of the captains and shipmasters on the vessels with alleged violations attended seminars conducted by personnel accredited by the AIDCP and the IATTC. In addition, crews from the vessels with alleged violations participated in workshops on IATTC conservation measures.

Venezuela recognizes that IATTC Resolution C-05-03 requires vessels to have onboard fins that total no more than 5 percent of the weight of sharks onboard, up to the first point of landing. When the fishing trips of the *Ventuari* and *Cayude* took place, Venezuela had not yet included a resolution prohibiting finning in its national regulations. Since then, however, Resolution No. 062, dated May 25, 2012, was published in the Official Gazette. The resolution contains regulations on the capture, exchange, distribution, commerce, and transport of sharks. It specifically prohibits finning of sharks, and states that fins must be naturally attached when landed, as well as the head, to allow for proper identification of the species.

Vessel-Specific Actions

Venezuela investigated the actions of the *Don Francesco*, *La Rosa Mística*, and *Taurus I*, which discarded salt bags or plastic trash at sea, and issued warnings to the vessels. Venezuela stated that, if the same activities take place in the future, monetary sanctions or suspension of fishing licenses could result from disciplinary proceedings.

Venezuela investigated the cases of the *Curimagua* and *Don Francesco* and determined that no infractions occurred, warranting no further administrative action. For the vessels that discarded tuna, Venezuela held seminars to train the captains and shipmasters on their responsibilities under the applicable regulations.

Certification

Venezuela passed a resolution governing the capture, exchange, distribution, commerce, and transport of sharks; issued warnings for discarding trash or plastic at sea; and investigated the two sea turtle release cases. It required training seminars be attended by each of its captains, shipmasters, and ship personnel who were on vessels with violations of IATTC conservation and management measures, including for discards of tuna. Thus NMFS has determined that Venezuela has taken appropriate corrective action to address the IUU fishing activities for which it was identified in the 2013 Report to Congress. Based on this finding, NMFS has made a positive certification determination for Venezuela.

V. State of Knowledge on the Status of International Living Marine Resources

Section 607 of the Moratorium Protection Act requires an accounting of the state of knowledge on the status of international living marine resources shared by the United States or subject to treaties or agreements to which the United States is a party, including a list of all fish stocks that are classified as overfished, overexploited, depleted, endangered, or threatened with extinction by any international or other authority charged with their management or conservation.[26] NMFS has updated the list that was cited in the 2013 Report to Congress, including links to the latest (as of fall 2014) status reviews of species. For each species, the table shows the status of each stock, the organization(s) that made the assessment, and applicable treaties. The revised list is available online at http://www.nmfs.noaa.gov/ia/iuu/msrapage/msra.html.

The list includes resources over which an international treaty or agreement, to which the United States is a party, has explicit conservation or management authority; has in place measures designed to control fishing mortality; or has directed the collection of fisheries data, including bycatch, to inform assessments of status. It also includes other resources shared by the United States, including U.S. territories, on which a directed fishery exists or which are taken as bycatch that are significant either in absolute numbers or because of the sensitivity of the international living marine resources, such as seabirds, sea turtles, marine mammals, or sharks, but which are not subject to an international treaty or agreement to which the United States is a party.

VI. International Actions to Address IUU Fishing

Global international organizations have acted in recent years to create many tools to combat IUU fishing and promote sustainable fisheries. This Part updates the descriptions of these activities in the 2013 Report to Congress.

Food and Agriculture Organization. Established in 1945, the FAO has a mandate to raise levels of nutrition and standards of living, improve agricultural productivity, and better the condition of rural populations. Today, the FAO is the largest autonomous agency within the UN system with 192 member countries plus the EU and one associate member (Faroe Islands). The FAO employs 1,600 professional staff and 2,000 general services staff.

The FAO's Committee on Fisheries (COFI), established in 1965, constitutes the only global intergovernmental forum other than UNGA where major international fisheries and aquaculture problems and issues are examined and recommendations addressed to governments, regional fisheries bodies, NGOs, fish workers, and the international community on a worldwide basis. COFI is also a forum in which global agreements, binding and non-binding, are negotiated.

Western Central Atlantic Fisheries Commission. WECAFC is a regional body established in 1973 under article VI(1) of the FAO Charter. As such, it does not have management authority for fisheries in the region, but helps members to promote effective conservation, management, and development of living marine resources in accordance with the FAO Code of Conduct for Responsible Fisheries. The Commission also addresses common problems of fisheries management and development faced by its members. WECAFC is composed of all 33 countries in the Wider Caribbean region and the EU.

WECAFC held its 15th biennial meeting in Port of Spain, Trinidad and Tobago, in March 2014. Although the status of many fisheries in this region warrants concern, there is clearly reason for optimism as evidenced by increased national and regional efforts for the management and conservation of certain species, many coordinated by WECAFC. In the midst of these accomplishments there is increased concern about IUU fishing in the Wider Caribbean. WECAFC created a new working group on IUU fishing, and convened a workshop on port State measures.

European Union–United States Bilateral Engagement. As two of the three top seafood importers in the world, the EU and the United States recognized their responsibility to protect the oceans' vital food and biodiversity resources in a historic statement signed in 2011 pledging bilateral cooperation to combat IUU fishing. Since then, the United States and the EU have worked together to support adoption of effective management measures in regional and international organizations, promote tools that prevent IUU operators from benefiting economically from their illegal activities, exchange information on IUU activities, and promote the sustainable use of fisheries resources while preserving marine biodiversity. U.S. and EU officials coordinated informally throughout 2013 and 2014 to continue their joint efforts to combat IUU fishing by identifying specific activities, dates, and points of contact, and through extensive discussion of regional and global fisheries issues. A staff-level working group met formally in 2014 to coordinate their efforts to combat IUU fishing.

In March 2014, the United States participated in a conference hosted by the EU on fishing capacity, where the link that often exists between overcapacity and IUU fishing was noted. The purpose of the conference was to reinvigorate the discussion of global fleet capacity at a high level and among States with a range of fishing interests. The primary outcome of the conference was a "Joint Statement on Efforts to Promote Sustainable Fishing Capacity Management on the Global Scale," which was signed by Colombia, the EU, Indonesia, Italy, Japan, the Philippines, and the United States. The statement identifies a number of activities to improve the management of fleet capacity.

The sections in this Part focus on particular approaches (such as port and flag State control measures) and specific tools (such as monitoring, vessel lists, and a global record of fishing vessels) that are being developed and implemented to deter IUU fishing activities.

A. Port State Measures

The reason IUU fishing continues despite decades of effort to curb the problem is the economic incentive that makes such activities cost-effective and financially viable for many fishermen and, indeed, investors. Removing or disrupting the economic drivers of IUU fishing promotes eradication of this global activity.

1. 2009 Agreement

On April 3, 2014, the Senate gave its advice and consent to ratification of the 2009 Agreement on Port State Measures to Prevent, Deter, and Eliminate Illegal, Unreported and Unregulated Fishing (PSMA), which requires parties to take actions to prevent IUU fish and fish products from entering the stream of commerce. The PSMA is the first binding global instrument focused specifically on combating IUU fishing. In recognition that all fish must pass through a port to get to market, the PSMA sets minimum standards for the conduct of port inspections and the training of inspectors. Parties must restrict port entry and access to port services to vessels that have engaged in IUU fishing, except when entry is allowed for the purpose of inspection, other enforcement actions, or circumstances of *force majeure*. The United States was a primary participant in the negotiation of the PSMA and one of the first countries to sign it.

Twenty-five nations must ratify the PSMA for it to enter into force; broad ratification is necessary for it to be effective in combating IUU fishing. Since the last report, another seven countries (Chile, Gabon, Mozambique, New Zealand, Oman, Seychelles, and Uruguay) have added their instruments of ratification, acceptance, approval, or accession to the first four. Timely ratification by the United States will enable NOAA and other agencies to continue actively encouraging ratification by other nations, but the United States can ratify only after implementing legislation is enacted. Bills to accomplish this were introduced in both the Senate (S. 267) and House (H.R. 69) during the 113th Congress, but were not passed.

2. RFMO Actions

While many RFMOs have adopted port State measures, frequently in conjunction with the measures that establish their IUU vessel lists, the PSMA, as a global, legally binding instrument, has the potential to fill many of the gaps that enable IUU fishermen to profit from their activities. As a result, many RFMOs are considering proposals to adopt or amend existing port State measures to be consistent with the minimum standards set forth in the PSMA.

ICCAT's Scheme for Minimum Standards for Inspection in Port entered into force in June 2013. It obliges port States to designate and publicize their ports where foreign fishing vessels may land or transship fish; calls for advance notice from such vessels seeking to enter those ports; provides that a port State must decide whether to grant entry to such vessels in light of the information received; and requires inspection, once in port, of at least 5 percent of landing or transshipment operations by foreign vessels equal to or greater than 12 meters length overall. At its annual meeting in 2013, ICCAT adopted forms for use in port inspections. In 2014 ICCAT adopted the U.S. proposal to further support effective implementation of the Scheme by establishing a special fund to provide technical assistance to port inspectors and other relevant enforcement personnel from developing coastal States.

In January 2014, SPRFMO adopted a CMM on minimum standards of inspection in port, which took effect January 1, 2015.

The United States continues to support adoption by the WCPFC of a port State scheme. Despite proposals by the EU and the Pacific Islands Forum Fisheries Agency to establish a port State inspection scheme at the regular sessions of the WCPFC in 2013 and 2014, no consensus was reached. Further discussion is expected at the WCPFC Technical and Compliance Committee and the annual meeting in 2015.

The IATTC has discussed port State measures since 2010. The most recent unsuccessful proposal, introduced at the 2014 annual meeting by the EU and supported by the United States, is modeled on the scheme adopted by ICCAT.

B. Market- and Trade-Related Measures

Trade- and market-related measures reduce opportunities for IUU fishing activities in a number of ways: by precluding or impeding access to markets for IUU fishing products in a manner consistent with international law; by tracking movements of fish products to identify those involved in harvesting, transshipping, and marketing of IUU catch; by monitoring changes in the pattern of trade to identify flag, port, and market States that can contribute to effective implementation of CMMs; and by improving information on fishing mortality. Successful market measures are often based on information gathered from trade-tracking programs – systems that can verify the origin, weight, and species composition of catch and indicate whether the catch was taken in accordance with the conservation and management regime in force.

NMFS is preparing a proposed rule to revise permitting requirements for those engaged in international trade of certain fish and fish products, and procedures for filing import and export documentation. This rule will help the United States to meet certain trade-related obligations that arise from participation in RFMOs. Specifically, NMFS intends to integrate the collection of trade documentation within the government-wide International Trade Data System and require electronic information collection through the automated portal maintained by U.S. Customs and Border Protection (CBP) of the Department of Homeland Security. Through this integration, NMFS would require annually renewable International Fisheries Trade Permits to import or export certain regulated seafood commodities. The permit pertains to fishery products that are subject to trade monitoring programs of RFMOs or subject to documentation requirements under domestic laws. These trade monitoring programs enable the United States to exclude products that do not meet the criteria for admissibility to U.S. markets, including products of IUU fishing activities. The proposed rule would consolidate existing international trade permits for seafood products, expand the scope of the permit requirement to additional seafood products, stipulate data and documentation that must be provided electronically to CBP, and address recordkeeping requirements in light of the proposed changes.

Early in 2014, NOAA signed a memorandum of understanding (MOU) with CBP to participate as a member agency in the Commercial Targeting and Analysis Center. At the Center, member agencies have direct access to a wide array of import processing, targeting, and law enforcement systems, as well as other members' data systems. NOAA analysts at the Center use the entire suite of systems and tools to monitor and analyze inbound trade to verify catch and trade documents, and use CBP's complex risk-based targeting system to identify suspect fisheries shipments before they arrive in the United States.

1. Global Forums

The United States routinely raises the need to prevent trade or import of IUU-caught fish and living marine resources in bilateral consultations and multilateral meetings and negotiations, as discussed throughout this report. In addition, the United States engages in the World Trade Organization and other trade-related bodies to eliminate subsidies that contribute to overfishing, overcapacity, or illegal fishing activities.

Convention on International Trade in Endangered Species

CITES is an international agreement among 180 member nations, with the purpose of ensuring that international trade in wild animals and plants does not threaten their survival. Species covered by CITES are listed in different appendices according to the level of protection needed. Any CITES party may add a native species to Appendix III unilaterally, provided that party has domestic laws to protect the species. The U.S. Fish and Wildlife Service is the lead agency with responsibility for implementing CITES in the United States, under the authority of the ESA. Based on its expertise, NOAA provides guidance on marine issues.

The 16th meeting of the Conference of the Parties to CITES (CoP16) took place in March 2013, in Bangkok, Thailand. Delegations from more than 170 member countries agreed to list several commercially harvested shark species in Appendix II of CITES for the first time: oceanic whitetip shark, three species of hammerhead sharks (scalloped, great, and smooth), and porbeagle shark (see Part IX.A for details). Also successful was a proposal by Ecuador to list manta rays in Appendix II; the proposal reflected concerns regarding the low rates of reproduction of manta rays and their high vulnerability due to growing international demand for their gill plates.

Another accomplishment, a priority for the United States, was adoption of a resolution regarding the provisions for trade in specimens taken "in the marine environment not under the jurisdiction of any State" that are listed in Appendix I or II of the Convention. Within CITES, trade in these specimens is referred to as "introduction from the sea." The United States has worked with other CITES parties for many years to resolve the permitting requirements for introduction from the sea. Within the framework of the resolution adopted at CoP16, if a vessel harvests CITES-listed specimens on the high seas and delivers them to the same country by which it is flagged, parties will treat the transaction as an introduction from the sea and issue an introduction from the sea certificate. Under this scenario, there is only one country involved in the trade. If more than one country is involved in the trade (that is, the vessel that harvests the specimens delivers them to a country other than the country by which it is flagged), CITES parties will treat the transaction as an export and require the issuance of an export permit by the country to which the harvesting vessel is flagged. The resolution adopted at CoP16, which also provides guidance in the case of chartering operations that are fishing on the high seas, will help provide greater certainty and consistency for transactions involving the several shark species that were listed in Appendix II at CoP16.

2. RFMO Actions

ICCAT annually reviews fishery-related activities in its Convention Area by members and nonmembers. In accordance with ICCAT's recommendation concerning trade measures, if an ICCAT member or non-member is found to be diminishing the effectiveness of ICCAT, the Commission may "identify" that member or non-member. ICCAT then sends a letter notifying the party of the identification, including the reasons for it, and asking the party to rectify the situation. Failure to rectify the identified activity may result in the imposition of penalties, such as quota reduction or, as a last resort, non-discriminatory trade restrictions. To date, trade-restrictive actions have been applied several times to non-members and once to an ICCAT member.

In 2013, ICCAT identified three members (Albania, Angola, and Honduras) and agreed to send 20 other members "letters of concern" noting specific issues that required correction. ICCAT's Compliance Committee also reviewed the actions of non-members, including those with cooperating status. Colombia's status as a cooperating non-member was revoked due to data reporting deficiencies and its failure to respond to its identification in 2012.

Also at the 2013 ICCAT annual meeting, the Compliance Committee conducted its first review under the Recommendation on Penalties Applicable in Case of Non-Fulfillment of Reporting Obligations. The Compliance Committee recommended action only in cases of contracting parties and cooperating non-contracting parties, entities, or fishing entities (CPCs) that submitted no catch data for any species. As a result, retention of any ICCAT species was prohibited in 2014 for the following members, pending receipt of catch data acceptable to the Standing Committee on Research and Statistics or a declaration that no catches of ICCAT species occurred during 2012: Albania, Equatorial Guinea, Gabon, Honduras, Republic of Guinea, and Sao Tome and Principe. ICCAT also agreed to request written verification of zero catches during 2012 from Nicaragua, Sierra Leone, and Syria.

At the 2014 ICCAT annual meeting, as a result of its review of CPC compliance, the Compliance Committee agreed that 25 CPCs will receive letters of concern. No identifications under ICCAT's trade measures instrument were made, as parties previously identified had made significant improvement in reporting and engagement, and other parties did not have non-compliance issues that rose to the level of identification.

The United States strongly supports moving toward electronic implementation of ICCAT catch and trade documentation programs. In 2013 and 2014, ICCAT continued work on the development of its electronic Bluefin Catch Document program, which is expected to enable verification of the legitimacy of products in near real-time; reduce the burden associated with a paper-based system on the seafood industry, governments, and the ICCAT Secretariat; and make it more difficult to falsify catch documents.

The WCPFC has discussed adoption of a CDS for several years, but with little progress. At its 2012 meeting, the Commission adopted terms of reference for a working group, which held its first meeting in 2013. The working group met again in 2014 and plans to produce an analysis of existing WCPFC member CDS-related initiatives for discussion in 2015.

At the October 2014 meeting of the IATTC, the Commission adopted a measure that included establishment of a CDS for Pacific bluefin tuna. The Commission for the Conservation of Southern Bluefin Tuna has had a CDS for that species since 2006. As a market State for southern bluefin tuna but a non-member of the Commission, the United States requires imports and exports of this tuna to be accompanied by a CDS document, but does not otherwise participate in the program. The United States will work with the organization to determine what information it can provide related to the CDS program, given domestic confidentiality limits.

CCAMLR instituted a CDS for trade in toothfish in 1999, but the system has become outdated. Between the 2013 and 2014 annual meetings, a CDS Review Panel conducted an independent review of CCAMLR's CDS. The Panel's efforts resulted in significant revisions to the CDS requirements to better enable monitoring of the movement of toothfish from point of landing through to market. A CDS Implementation Panel will further review the CDS to improve transparency with regard to transshipments; integrate the CDS with other related databases, such as VMS and reported catch data; and upgrade the e-CDS system through a new platform.

C. Monitoring, Control, and Surveillance

1. Information Sharing and Coordination

International information sharing and coordination aimed at deterring IUU fishing takes many forms: cooperation among national authorities to enforce regional and global measures, assistance to developing nations in protecting their own natural resources, and adoption of RFMO procedures to facilitate information sharing on enforcement matters.

As an example, NOAA and the USCG work closely with enforcement agencies from Canada, Japan, the Republic of Korea, and the Russian Federation to enforce the North Pacific Anadromous Fisheries Commission (NPAFC) prohibition on directed fishing for anadromous stocks in the high seas areas of the North Pacific Ocean. NPAFC enforcement activities contribute significantly to implementation of the UN global moratorium on large-scale high seas driftnet fishing. NPAFC members coordinate multilateral air and surface patrols to utilize enforcement resources more efficiently. Each spring the parties discuss current enforcement efforts, coordination of enforcement plans, and resource sharing for the remainder of the calendar year.

The overall reduction in sightings of vessels engaged in illegal fishing activity in the North Pacific in recent years testifies to the effectiveness of the Commission's cooperative enforcement model. In 2013, no vessels were observed or apprehended for illegal salmon fishing in the NPAFC Convention Area. The combined monitoring activities in 2013 by NPAFC-related enforcement agencies included more than 120 ship patrol days, more than 498 aerial patrol hours, and satellite support. Members collaborated through joint ship patrols, exchanges of personnel in the air and ship patrols of member countries, and regular conference calls.

In 2014, parties conducted at least 100 ship patrol days and 440 aerial patrol hours in the Convention Area. On May 18, 2014, the USCG Cutter *Morgenthau* intercepted two suspected IUU fish transshipment vessels about 35 nautical miles outside the Russian EEZ – the *Sungari*, flying a Sierra Leone flag, and the *Sovereign*, flying a Russian flag. The Government of Sierra Leone denied that the *Sungari* was registered in Sierra Leone. While the investigation was underway, the *Sungari* changed its name to *Stellar* and raised the Cambodian flag.[27]

The *Morgenthau* broke off its investigation of the two vessels on May 20, 2014, to pursue a suspected high seas driftnet fishing vessel, the *Yin Yuan*, located by a Canadian CP-140 patrol aircraft about 250 nautical miles south of the *Morgenthau*'s position. The *Yin Yuan* was flying a Japanese flag; however, Japanese fisheries observers on board the CP-140 stated that the vessel was not a Japanese-registered vessel. When the *Yin Yuan* became aware of the aircraft, a crew member struck the Japanese flag. A large net, net tube, and net spreader were clearly visible on deck, and a large number of buoys were later located throughout the vessel – all typical of a driftnet vessel.

The *Morgenthau* pursued the *Yin Yuan* for several days. On May 25, when questioned, the ship's master made a verbal claim of Chinese registry. Chinese shipriders aboard the *Morgenthau* directed the *Yin Yuan* to heave-to for inspection. During an initial interview, the master said the vessel was not fishing, only transporting gear to another fishing vessel. When shown pictures of the *Yin Yuan*'s nets, net tube, and net spreader on board, he admitted to having dumped them overboard the night of May 23. He added that 3.3 kilometers of driftnet had also been dumped over the side. The USCG boarding team found a container holding approximately half a ton of salmon. On June 3, 2014, the *Morgenthau* escorted the *Yin Yuan* to a location in the East China Sea and transferred custody of the vessel to the Chinese Coast Guard for prosecution.

This case demonstrates the utility of the Chinese shiprider program, established by an MOU signed in 1993. For more than two decades, the USCG, in conjunction with NOAA, has embarked members of China's Fisheries Law Enforcement Command on USCG assets patrolling the highest threat areas in the North Pacific Ocean for illegal high seas driftnet fishing. These patrols support the UN global moratorium and provisions of the Convention for the Conservation of Anadromous Stocks in the North Pacific Ocean. They also enable China to more successfully enforce domestic laws that prohibit high seas driftnet fishing by Chinese-flagged vessels in the North Pacific, and effectively expand the jurisdictional reach of both enforcement agencies. The MOU was renewed in late 2014 for another five years, until December 31, 2019.

During 2013 and 2014, the USCG participated in NAFO at-sea inspections by deploying USCG boarding officers as shipriders on Canadian Coast Guard vessels on three joint patrols. Sixteen vessels were inspected, with no violations detected.

The North Atlantic Salmon Conservation Organization (NASCO) monitors for possible IUU activity for Atlantic salmon on the high seas, especially in the area north of the Faroe Islands, through cooperation with the Norwegian and Icelandic Coast Guards. NASCO's independent performance review indicated that, given the limited nature of this surveillance in recent years, other methods should be explored to better assess whether IUU fishing of Atlantic salmon may be occurring in international waters. NASCO has requested the cooperation of NEAFC and NAFO in reporting on suspected

IUU fishing activities for salmon that may be detected in their MCS operations.

In 2013 and 2014, the USCG patrolled the high seas portions of the WCPFC Convention Area, pursuant to the WCPFC High Seas Boarding and Inspection Procedures. The majority of fishing vessels inspected were operating in compliance with WCPFC conservation measures; only a few minor violations were identified.

The IOTC adopted in 2012, and updated in 2013, a resolution requiring members to report information on access agreements. ICCAT's similar measure, adopted in 2011, was revised in 2014 to require additional information reporting.

2. International MCS Network and INTERPOL

The United States is one of the founding members of the International Monitoring, Control and Surveillance Network (IMCS Network). NOAA participates in the IMCS Network as one mechanism for sharing information and experience with fisheries law enforcement professionals from other countries. A voluntary organization of more than 50 member countries, the IMCS Network is the only international organization specifically dedicated to the training and support of fisheries law enforcement practitioners directly engaged in the battle against IUU fishing. NOAA contributes funds and technical support to the Network. A NOAA enforcement officer currently serves as the vice chair. The IMCS Network conducted its fourth enforcement training workshop in Costa Rica in February 2014. Additional information on the IMCS Network is available at www.imcsnet.org.

The INTERPOL Fisheries Crime Working Group met in 2013 and 2014 in Lyon, France. For the past two years, a NOAA official has served as vice chair of the group. At the most recent meeting, 70 representatives from 40 member countries addressed the use of INTERPOL's tools and services for fisheries law enforcement, operational planning and training, and the global effect of criminal activities at sea. Two specialized sub-groups concentrate on capacity building and intelligence. The Working Group developed the first worldwide INTERPOL Purple Notice for a CCAMLR IUU-listed vessel, at the request of Norway in September 2013.

NOAA participates in INTERPOL Washington, the U.S. National Central Bureau, as program manager of the Environmental Crimes Unit. INTERPOL Washington issued its first Purple Notice in June 2014, making all INTERPOL members aware of the activities of the *Stellar/Sungari* described in Part VI.C.1. INTERPOL Washington has also issued five Red Notices for the arrest of fugitives wanted for maritime environmental crimes. A new INTERPOL operation beginning in November 2014, involving more than 130 Red Notices, has already resulted in four arrests.

3. IUU Vessel Lists

ICCAT reviews its IUU vessel list annually. Members and cooperating non-members must take necessary measures not to support listed vessels, including by prohibiting imports, landings, or transshipments of ICCAT species. The IATTC and IOTC conduct similar reviews of their IUU vessel lists. At its 2014 meeting, ICCAT adopted guidelines on implementation of the requirement to cross-list vessels appearing on other tuna RFMO IUU vessel lists with the ICCAT IUU vessel list. At the July 2014 meeting of the IATTC, the United States co-sponsored a proposal to clarify the procedures to add or remove a vessel to the IUU list; however, this proposal was not adopted.

CCAMLR did not consider any new vessels for its two IUU vessel lists at the 2013 or 2014 meeting. A request to remove the *Tchaw* (currently a stateless vessel) from one list was not supported by sufficient information. One vessel, the *Tiantai*, was removed from an IUU list on the basis of its presumed sinking in the Convention Area. CCAMLR expects increased cooperation with INTERPOL in gathering information on the locations and activities of vessels on its IUU list.

The WCPFC received no requests to remove any vessel from its 2013 or 2014 IUU Vessel List, nor did it receive nominations for vessels to be added to the list. As a result, the 2015 IUU Vessel List remains unchanged, with three vessels.

NAFO maintains a list of vessels that have conducted IUU fishing in its Regulatory Area, and shares IUU vessel sightings with other RFMOs with interests in the region. NAFO collaborates particularly closely with NEAFC, as the two RFMOs are adjacent, share much of the same membership, and manage groundfish stocks that are susceptible to IUU fishing by the same vessels. NAFO and NEAFC recognize each other's vessel list, which allows membership from both organizations to act in concert to restrict port access by IUU-listed vessels and to delist vessels as appropriate.

SPRFMO adopted measures to establish a list of vessels presumed to have carried out IUU fishing activities in its Convention Area in 2013, and to require the collection, reporting, verification, and exchange of data. In 2014, SPRFMO adopted additional information sharing and coordination measures, including requirements for a record of vessels authorized to fish in the Convention Area.

4. Global Record of Fishing Vessels

The FAO initiative to compile a Global Record of Fishing Vessels, Refrigerated Transport Vessels and Supply Vessels is intended to provide a tool to prevent, deter, and eliminate IUU fishing and related activities. A global database, where information from many sources will be gathered, will make it more difficult and expensive for vessels and companies acting illegally to do business. Eventually, all vessels 10 gross tons or 10 gross registered tons or more, or 12 meters or more, will be included (an estimated 725,600 vessels). During the first phase, around 185,600 of the largest vessels (100 gross tons or 100 gross registered tons or more, or 24 meters or more) will enter the record.

The June 2014 meeting of COFI took a number of important steps to reinvigorate the Global Record initiative. It endorsed a plan to complete the first phase in five years by adding to an existing database rather than creating a stand-alone database. COFI resolved that the IMO numbering scheme would be the Unique Vessel Identifier system for those vessels 100 gross tons and over. In addition, the Government of Spain pledged 250,000€ toward the costs of the first phase. NMFS is also a financial supporter of the first phase. The United States proposed, and COFI agreed, to establish a Global Record Steering Committee; the United States has agreed to serve on that Committee.

Issuance of IMO numbers to fishing vessels is a critical step in combating IUU fishing. The seven-digit number, once assigned to a vessel, remains unchanged upon transfer of the ship to other owners or flags or changes in name. Assignment of an IMO number reduces the potential for IUU vessels to evade enforcement and prosecution through transfer to flags of convenience. The IMO, with U.S. support, in December 2013 amended one of its resolutions to allow fishing vessels to obtain IMO numbers.

Since 2013 and as a direct result of U.S. leadership, all tuna RFMOs to which the United States is a party (IATTC, ICCAT, and WCPFC), as well as NAFO and CCAMLR, have adopted binding measures requiring all eligible vessels to have an IMO number. In the case of ICCAT, if a vessel included on ICCAT's record of vessels authorized to fish for tuna and tuna-like species in the Convention Area is unable to obtain an IMO number, the flag State must provide an explanation of this situation to ICCAT. SPRFMO's requirement for IMO numbers will take effect January 1, 2016.

An example of a regional approach to a record of authorized vessels comes from the WCPFC, which in 2012 implemented a CMM establishing an interim register of non-member carrier and bunker vessels. The interim list expired in 2013, at which time non-member carrier and bunker vessels were no longer allowed to operate in the WCPFC Area. This ban provided an incentive for carriers and bunkers to flag to WCPFC members, cooperating non-members, and participating territories. This action also encouraged flag States of carriers and bunkers to become cooperating non-members of the WCPFC. In 2013, after more than two years of negotiation, the WCPFC adopted a CMM to establish standards, specifications, and procedures for the WCPFC Record of Fishing Vessels. These are aimed at making the record as reliable as possible.

5. Remote Sensing Technology, Observers, and Inspections

NMFS promotes the adoption of VMS provisions by RFMOs and flag States. Currently NMFS monitors nearly 4,500 fishing vessels required to carry VMS equipment in more than 20 fisheries and permit areas, including several treaty or international agreement areas. U.S. enforcement personnel assist in crafting RFMO conservation measures requiring VMS use, and provide training on the use of VMS in fisheries enforcement.

The IATTC took an important step toward improving the amount and quality of scientific data provided to the Commission by mandating a minimum of 5 percent observer coverage on all longline vessels greater than 20 meters length overall, effective January 1, 2013. The first review of this measure, scheduled for 2014, was not completed. At the IATTC compliance committee meeting, various members discussed their difficulties in meeting 5 percent observer coverage. The IOTC adopted a similar measure, requiring 5 percent observer coverage for each gear type on all fleets for vessels over 24 meters, and for vessels under 24 meters that fish outside their own EEZs, effective January 2013. A review of the implementation of the IOTC measure in 2014 indicates that 12 IOTC countries have deployed observers; however, the rate of required observer coverage has not yet been met in most fisheries.

The IATTC adopted an update to its VMS measure in 2014. The revised measure, which will take effect in January 2016, requires more frequent transmission of vessel position reports from every six hours to every four hours for longline vessels and every two hours for all other fishing vessels.

At its 2013 annual meeting, ICCAT strengthened a conservation measure to require chartered vessels to have observer coverage of at least 10 percent of fishing effort. The other minimum standards established for scientific observers now apply to chartered vessels. In addition, ICCAT has refined its requirements, including a standardized protocol and common procedures, for the use of stereoscopic cameras in eastern Atlantic/Mediterranean bluefin caging operations. Beginning in 2015, these multi-lens cameras will be required in all caging operations to ensure accurate estimates of the number and size of caged fish, along with validation and verification procedures and the provision of video footage to observers. At its 2014 annual meeting, ICCAT also revised its minimum standards for VMS to require more frequent transmission of vessel position reports (from every six to every four hours) and the transmission of position reports to the coastal State when a flag vessel from another State or entity is fishing in the waters under the jurisdiction of that coastal State.

The WCPFC Convention requires that all vessels fishing for highly migratory fish stocks on the high seas in its Convention Area participate in a VMS operated by the Commission. The system has been largely operational for the past few years, and was fully applied to all vessels by the end of 2013. At the request of any coastal State member, the WCPFC's VMS may also include waters under the member's national jurisdiction. This feature, unique to WCPFC, greatly expands the utility of the VMS and provides coastal States with enhanced ability to detect foreign vessels fishing illegally in their waters. Ten members, including the United States, have chosen to include waters under their jurisdiction in the Commission VMS.

In 2013, a CCAMLR working group developed a set of technical requirements and other enhancements to modernize and improve the functionality of the CCAMLR VMS. At the 2014 meeting, the United States proposed requiring real-time reporting of positions directly to the Secretariat as well as the flag State, reporting of positions every hour, reporting for the duration (port to port) of any trips during which the vessel operates in the Convention Area, and other changes. While some members were supportive of the changes, other members stated that they needed additional time to consider them.

Also at the 2014 CCAMLR meeting, France and the Secretariat proposed a six-month pilot initiative using satellite-derived imagery to estimate IUU fishing vessel presence in a high-seas area of the Convention Area. The pilot may yield information on vessel activity and fishing effort but would not provide information on catches or species composition. The EU intends to provide 375,000€, subject to confirmation of fund availability. Australia and the United States expressed interest in collaborating on this pilot.

NAFO established a compliance-based observer program in 1998, which requires all contracting party vessels to carry at least one observer when operating in its Regulatory Area, as well as mandatory use of VMS. In 2013 and 2014, NAFO took further steps to improve the accuracy of catch reporting data and to enhance its utility. At its 2014 meeting, NAFO approved further meetings of existing working groups on catch reporting and bycatch.

In 2014, SPRFMO adopted a measure to establish standards for the nature and specification of a planned but not yet operational Commission VMS.

D. Flag State Responsibilities

IUU fishing is exacerbated or even inadvertently encouraged by irresponsible flag States – in particular, those States that allow vessels to fly their flags without any capability or effort to monitor and control the operations of those vessels. In response to the perceived failings of several flag States, the FAO worked for several years to develop criteria for assessing the performance of flag States.

On February 7, 2013, the FAO Technical Consultation on Flag State Performance successfully completed negotiations on the Voluntary Guidelines for Flag State Performance. Going into the week's session, there were three major outstanding issues: the scope of the guidelines, the question of whether external entities could initiate a performance assessment of a flag State, and the consequences of a performance assessment. The meeting concluded that the scope extended to high seas areas; the guidelines might also apply to fishing and related activities within the national jurisdiction of the flag State or of a coastal State, upon their respective consent. The meeting decided that the guidelines would not provide for third-party initiation of a performance assessment, but recognized that anyone could apply them (although it is hard to imagine a useful flag State performance assessment without the cooperation of that flag State). Finally, the meeting agreed to include a U.S.-proposed paragraph addressing the possible consequences of positive and negative flag State performance assessments. The potential sanctions in the International Plan of Action to Prevent, Deter and Eliminate IUU Fishing, particularly its section on "internationally agreed market-related measures," are incorporated by reference.

After the session, representatives of the delegations that had worked most closely with the United States agreed that the outcome, while not ideal, was close to the best that could have been achieved, and that, after more than a decade of designing new tools to combat IUU fishing, FAO members had finally taken steps to get at the heart of the problem – the obligations and duties of flag States.

Since COFI adopted the guidelines on June 11, 2014, Canada has informed the United States of its intent to perform a self-assessment to demonstrate to other countries how easily and productively the guidelines can be used. The United States has been consulting with other governments on the suggestion that RFMOs require a self-assessment by countries applying for non-member status in their organizations.

E. Destructive Fishing Practices and Vulnerable Marine Ecosystems

The United States and the international community have taken a number of actions in recent years to mitigate the adverse impacts from fishing on VMEs (see Part II.A for the statutory definition). These actions built on guidance from several UNGA resolutions on sustainable fisheries, specifically Resolutions 61/105, 64/72, and 66/88.

In 2013 and 2014, the United States supported the FAO in its efforts to improve global implementation of the UNGA mandate to protect VMEs, including through a project funded by the Global Environment Facility. In addition, NOAA experts participated in FAO workshops to identify VMEs in the North Pacific and the Central Atlantic. The results of these meetings will enhance shared understanding of the location of VME indicator species and the potential impacts of fishing activities on them.

Since 2008, when the resolution came fully into effect, the five RFMOs that actively manage bottom fishing have implemented measures to comply with Resolution 61/105. (As explained in Part II.A, the Southeast Indian Ocean Fisheries Agreement has not implemented such measures, but NMFS has concluded that all fishing activities in the area covered by that Agreement are consistent with Resolution 61/105.)

The United States has taken a strong role in NAFO to protect VMEs. Since 2006, NAFO has closed more than 360,000 square kilometers to bottom fishing. In 2013, NAFO continued to enhance this protection by closing a number of VME areas within its Regulatory Area to all fishing. In 2014, NAFO extended these closures until 2020 and added two new areas. NAFO is discussing possible impacts of mid-water trawls on or near seamounts, and the utility of encounter thresholds within its "fishing footprint." NAFO is also preparing for a comprehensive re-assessment of all its bottom fisheries.

Five notifications of encounters with potential VMEs during exploratory bottom fisheries in the Convention Area were submitted to CCAMLR during the 2012–2013 season, with one new VME risk area declared in a Subarea in 2013. During the 2013–2014 fishing season, one VME indicator notification was submitted, adding to the 155 notifications from exploratory bottom fisheries. These notifications led to the declaration of 64 VME risk areas closed to fishing since the adoption of the conservation measure in 2008. CCAMLR intends to review its VME-related conservation measures, to evaluate their effectiveness in protecting VMEs from significant adverse impacts.

SPRFMO adopted a measure for the management of bottom fishing in the SPRFMO Convention Area in 2014. The measure sets forth specific requirements for bottom fishing activities, including catch limits, requirements to cease bottom fishing activities within five nautical miles of any site in the Convention Area where evidence of a VME is encountered above certain threshold levels, and closing areas where VMEs are known to occur or are likely to occur unless bottom fishing would not have significant adverse impacts on VMEs.

VII. Strengthening Fisheries Management Organizations to End IUU Fishing Activities

The United States aggressively combats IUU fishing through global bodies such as UNGA and the FAO, regionally through RFMOs, and through bilateral engagement. The United States is a member of numerous multilateral RFMOs, in addition to many global and bilateral agreements and arrangements.

In recent years, the international community increasingly recognized that successful action against IUU fishing activities and related problems would require strengthening existing regional fisheries institutions and creating new RFMOs to manage previously unregulated ocean areas. The United States has been a major force in these efforts, as discussed below. This Part highlights the establishment of new organizations and the enhancement of existing ones in ways that induce their members to be more accountable, and influence non-members to be more cooperative, in managing fisheries on a sustainable basis.

A. Establishing New RFMOs

Due to the efforts of the United States and many others, the number of RFMOs continues to expand. This section describes developments in nascent RFMOs since the 2013 Report to Congress.

South Pacific Regional Fisheries Management Organization (SPRFMO)

The Convention on the Conservation and Management of the High Seas Fishery Resources of the South Pacific Ocean entered into force on August 24, 2012. This organization fills a gap in the international fisheries management regime in the South Pacific, and responds to recent calls from the UN and elsewhere to take urgent action with regard to the impacts of destructive fishing practices on high seas VMEs. The main fisheries currently addressed by SPRFMO are pelagic fisheries for jack mackerel and bottom fisheries for species such as orange roughy. During the course of the negotiations, the participants agreed to non-binding interim CMMs covering both pelagic and bottom fisheries.

The first two meetings of SPRFMO took place in January 2013 and January 2014. The United States is a signatory to the Convention and a cooperating non-contracting party to SPRFMO. The Senate provided its advice and consent to ratification of the Convention on April 3, 2014; the Administration submitted proposed legislation to Congress in June 2014 that would allow the United States to become a party to the organization.

North Pacific Fisheries Commission (NPFC)

The goal of the NPFC is to ensure the long-term conservation and sustainable use of the fisheries resources in the North Pacific Ocean, while also protecting the marine ecosystems in which these resources occur. The Convention on the Conservation and Management of High Seas Fisheries Resources in the North Pacific Ocean was opened for signature in April 2012, with the United States the first signatory. The Convention will enter into force when four parties have deposited their instruments of ratification; as of August 2014, three parties have done so. The Senate gave its advice and consent to U.S. ratification on April 3, 2014, which can occur only after Congress passes implementing legislation.

Six sessions of the Preparatory Conference have addressed the administrative and budgetary issues of setting up a Secretariat, the selection of Japan as the Secretariat host nation, and establishing MCS procedures. The Convention establishes a management framework for all fisheries not already covered under existing international management instruments, with a particular focus on bottom fisheries, across the high seas areas of the North Pacific. Interim measures have been agreed with regard to bottom fishing, including the compilation, analysis, and exchange of data on bottom fishing in the region, and steps to protect VMEs from the impacts of bottom fisheries.

B. Strengthening Existing RFMOs

In addition to establishing new RFMOs, the United States has pushed for improved governance systems in existing RFMOs to bring them into closer conformity with the provisions of the United Nations Fish Stocks Agreement (UNFSA). Some RFMOs have been updated through renegotiation of their underlying agreements or negotiation of new protocols. Others are finding ways to improve management and compliance without renegotiating their agreements. This section reviews developments in existing RFMOs since the 2013 Report to Congress.

1. Renegotiation or Amendment of Underlying Agreements

After several years of discussion, informed by an independent performance review in 2008, ICCAT agreed at its 2012 meeting to launch a process to develop Convention amendments concerning scope, decision-making procedures, and non-party participation, among other subjects. A working group met in 2013 and 2014 to exchange ideas and has begun the process of developing draft Convention text. The United States will host the third and expected final meeting of the working group in May 2015.

U.S. officials were heavily involved in negotiating an agreement to update and modernize the guiding principles, mandate, and functions of the IATTC, a body established in 1949 to manage tuna fisheries in the Eastern Pacific Ocean (EPO). The new agreement – the Antigua Convention – entered into force on August 27, 2010. The Senate in 2005 provided its advice and consent to U.S. ratification, which is subject to the passage of legislation to clarify U.S. authorities to implement the Antigua Convention. In the 113th Congress, implementing legislation was introduced as Title IV of S. 269 and Title II of H.R. 69. The Senate Commerce Committee reported favorably on S. 269 on December 17, 2013; H.R. 69 was ordered to be reported by the House Natural Resources Committee on September 18, 2014. Neither bill was passed.

NAFO in 2007 adopted comprehensive amendments to its Convention that will enter into force once nine contracting parties deposit their instruments of ratification. To date, Canada, Cuba, the EU, Norway, and the Russian Federation have done so. The Senate gave its advice and consent to the amendments on April 3, 2014; but, again, ratification depends upon enactment of implementing legislation to revise the Northwest Atlantic Fisheries Convention Act of 1995.

2. Performance Reviews

Many RFMOs have undertaken performance reviews to bolster their organizations. The Review Conference on the Fish Stocks Agreement at its meeting in May 2010 urged all RFMOs that had not undertaken performance reviews, including some element of independent evaluation, to do so no later than 2012.

The SPRFMO Convention includes a requirement for a performance review every five years (Article 30).

At its 2014 annual meeting, ICCAT established an ad hoc working group that is charged with preparing proposed terms of reference for the next performance review of ICCAT.

As described in the 2009 Report to Congress, NASCO embarked on a comprehensive performance review in 2004. At the 2012 meeting, a panel of independent experts presented its recommendations, including that NASCO should explore whether and how to make more of its decisions binding. A meeting of the NASCO parties in February 2013 developed a prioritized plan of action, which was agreed at the 2013 annual meeting. Key recommendations to strengthen the organization included improving transparency and accountability, particularly by enhancing its procedures for members to report on and review implementation of agreements, and increasing outreach to other RFMOs on MCS issues. In light of these and other actions, NASCO declined to embark on a process to amend its Convention.

During 2013 and 2014, NAFO continued to implement the recommendations of its 2011 Performance Review Panel. A report by the Secretariat in 2014 showed considerable progress.

The CCAMLR Secretariat summarized recommendations from its 2008 performance review and, at the 2014 annual meeting, presented a range of options relating to the timing, scope, procedures, and possible development of terms of reference for a second performance review. The Commission intends to consider terms of reference for the review at the next annual meeting, with the EU preparing a draft that it will make available for comment during the intersessional period.

An NPAFC review panel consisting of outside experts and chairs of the Commission's standing committees presented a report on its performance review to the annual meeting in 2010. The reviewers had a total of 54 recommendations for the Commission, most of which were procedural in nature. The Commission has now completed 52 of the tasks.

In 2011, the WCPFC received a set of 79 recommendations from an independent performance review conducted by representatives of members and independent experts. At the 2013 meeting of the Commission, the Secretariat presented a paper containing a detailed matrix of recommendations from the performance review, sorted by subsidiary body to which each recommendation relates and including progress in addressing each issue. Subsidiary bodies continue to consider relevant recommendations. The Secretariat circulated a revised version of the matrix, removing all recommendations that have been addressed and completed; however, no decisions were made on the remaining recommendations at the 2014 Commission meeting.

The IOTC has agreed to undertake a second performance review of its operations, with terms of reference setting the objective of evaluating progress made on recommendations arising from the first performance review.

The IATTC, at its October 2014 meeting, adopted a proposal for a performance review that focuses on a review of the Secretariat, with some provisions to review the performance of the Commission. This review is to be completed by the 2016 annual meeting at the latest.

3. Bolstering Responsibilities of Members and Non-Members

In 2013, CCAMLR for the first time implemented a compliance evaluation procedure, and adopted a report by its Standing Committee on Implementation and Compliance that identified instances of non-compliance. The report revealed gaps in relation to technical aspects of the VMS measure and identified other CMMs that need revision. The 2014 compliance evaluation categorized more instances of non-compliance and resulted in amendment of several measures.

In 2011, the WCPFC initiated a trial scheme for compliance monitoring. The CMS continues to improve after four years of implementation. In 2014, the United States helped develop a Secretariat proposal for a more robust, multi-year CMS that would better define potential consequences for non-compliance. Commission members were unable to reach consensus on the multi-year proposal, so the existing measure will remain in effect for 2015, with the addition of a list of priority obligations to replace those in the existing measure.

Based on recommendations from its internal and external performance review processes, NASCO's action plan includes establishing an annual review of actions taken by parties to implement NASCO agreements and to fulfill their treaty obligations. The objective is to help ensure accountability by the parties and create an incentive for further action where needed. In 2014, the organization streamlined its work to allow more time for discussion of annual progress reports from each jurisdiction and to hold a special session on single- and mixed-stock fisheries, especially those involving stocks below their conservation limits. The theme-based special session clearly demonstrated where parties had or had not fulfilled their commitments. For some, the process led to a commitment to improve implementation.

In 2014, the IATTC Secretariat began distributing a list of possible infractions and corresponding explanations, which could include results of investigations, by each member and cooperating non-member.

In 2014, SPRFMO's Compliance and Technical Committee discussed an informal paper on the development of a CMS. The Committee will continue to work on this matter, with the EU leading the effort.

4. Steps to Enhance Participation by Non-Members

Consistent with the provision of the UNFSA relating to the duty of non-members to cooperate in the conservation and management of fish stocks, RFMOs are working toward enhanced participation by non-members in their organizations.

The IATTC has four cooperating non-parties: Bolivia, Honduras, Indonesia, and Liberia. The IOTC has three cooperating non-contracting parties: Djibouti, Senegal, and South Africa.

In CCAMLR, States that have acceded to the Convention but have not applied for membership in the Commission are obligated to abide by all the conservation measures adopted by the Commission and may not fish in the Convention Area. Currently there are 11 such nonmembers. In addition, any NCP may cooperate with CCAMLR by participating in its CDS; Seychelles is the only country with that status. Singapore has been provided limited access to the CDS for monitoring of its toothfish trade.

Following a substantial revision of the WCPFC measure pertaining to cooperating non-member status in 2008, the Commission again amended these provisions in 2009 to add a requirement that an applicant for this status must make financial contributions commensurate with what it would be assessed should it become a contracting party. In 2014, the WCPFC approved this status for eight applicants: Ecuador, El Salvador, Liberia, Mexico, Panama, Senegal, Thailand, and Vietnam. Senegal's acceptance was conditioned upon a financial contribution.

Currently, ICCAT has five cooperating non-members. At the 2014 annual meeting, ICCAT renewed the cooperating status of Bolivia, Chinese Taipei, El Salvador, and Suriname, and reinstated that status for Guyana. Through the process to amend the ICCAT Convention, ICCAT is considering approaches to enhance the participation of non-members in Commission activities.

SPRFMO adopted rules for cooperating non-contracting parties in 2013. The United States is currently in that status, along with Colombia, Ecuador, Liberia, Panama, and Peru.

5. Steps to Improve Cooperation and Coordination

Representatives of RFMOs are working to improve cooperation and coordination among themselves, particularly for those operating in the same region or managing highly migratory species.

The Kobe Process refers to such efforts undertaken by the five tuna RFMOs. The first meeting occurred in Kobe, Japan, in 2007; the second was hosted by the European Community, in San Sebastian, Spain, in summer 2009. The United States hosted Kobe III in La Jolla in July 2011, where principles were forwarded to the five tuna RFMOs as guidance in harmonizing criteria and processes, so that each tuna RFMO list could include IUU vessels identified by the others. A steering committee of the RFMO chairs and vice chairs convened at the 2012 and 2014 COFI meetings. In 2014, steering committee members reported on progress in implementing past Kobe recommendations and agreed on further coordination among the five tuna RFMOs.

As noted previously, NASCO is seeking enhanced cooperation with NAFO and NEAFC on MCS matters to improve detection of any IUU fishing for Atlantic salmon. NASCO also adopted an MOU with the Convention for the Protection of the Marine Environment of the North-East Atlantic to improve information sharing and cooperation with regard to matters that can impact the conservation of wild Atlantic salmon.

VIII. International Efforts to Reduce Impacts of Fishing on PLMRs

The United States continues to work actively within the international community to promote measures that will protect and conserve PLMRs from bycatch or other harmful activities. U.S. bilateral and multilateral efforts include direct advocacy as well as training and other assistance. To date, U.S. efforts and RFMO actions concerning PLMRs have generally concentrated on the impacts of fishing on sea turtles, sharks (see Part IX), dolphins, and some other marine mammals. This Part describes the actions taken by international bodies with regard to these PLMRs, and U.S. involvement in those actions.

A. Global Forums

United Nations General Assembly (UNGA)

As a direct result of U.S. leadership, the UNGA 2011 Sustainable Fisheries Resolution called for States and RFMOs to establish or strengthen existing data collection programs for the bycatch of marine mammals, sea turtles, seabirds, and sharks, in addition to supporting research on and development of appropriately selective gears. The 2013 Sustainable Fisheries Resolution calls on States and RFMOs to coordinate in developing and implementing clear and standardized bycatch data collection and reporting protocols for non-target species. The 2014 Resolution calls on States to encourage full utilization of sharks caught in sustainably managed fisheries and also notes the continuing practice of fins being removed from sharks while the rest of the carcass is discarded at sea.

International Whaling Commission (IWC)

An IWC working group, including Australia, Brazil, Canada, Mexico, New Zealand, South Africa, and the United States, upgrades response and reporting capabilities in developing countries to address entanglements of large whales in fishing gear. An IWC Secretariat staff member conducts entanglement response training at the request of member nations. The first entanglement response training workshop was held in March 2012. Since then, the initiative has reached more than 500 conservationists, scientists, and government representatives from more than 20 countries. In 2013 and 2014, the IWC held three training workshops in many countries, including nations in

South America, the South Pacific, and the Caribbean. In 2014, the team continued to develop standardized best practices and consulted on difficult entanglement cases.

Convention on the Conservation of Migratory Species of Wild Animals (CMS)

Also known as the Bonn Convention, the CMS aims to conserve terrestrial, marine, and avian migratory species throughout their range. The United States is not a party to CMS, but attends meetings as a non-party. In November 2014, the 11th Meeting of the Conference of the Parties to CMS adopted 31 proposals to add species to the Convention's two appendices, including 21 shark, ray, and sawfish species. The CMS parties also adopted several resolutions, one on the conservation of sharks and rays and another on a single action plan for the loggerhead turtle in the South Pacific Ocean.

B. RFMOs

During its 2014 annual meeting, NAFO adopted a U.S. proposal on reporting bycatch and discards. The measure extended a working group focusing on bycatch; initiated a review of existing bycatch data mechanisms and recommendations to improve the collection, reporting, and use of bycatch data; and a call to more broadly address ecosystem considerations through a comprehensive strategy on all bycatch and discards.

ICCAT directs CPCs to require collection of bycatch and discard data through existing logbook and observer programs, and to report these data in a standardized electronic form established in 2014. ICCAT has also adopted minimum standards for observer coverage, including 5 percent minimum coverage for pelagic longline, purse seine, and baitboat fisheries. For artisanal fisheries that are not subject to ICCAT's standards and requirements, CPCs are required to collect bycatch data through other means and to describe their efforts in annual reports. Since 2012, CPCs are also required to report on steps taken domestically to mitigate bycatch and reduce discards.

Under its Convention, the WCPFC is to adopt measures to minimize waste, discards, catch by lost or abandoned gear, catch of non-target species (both fish and non-fish), and impacts on associated or dependent species (particularly endangered species). Another mandate is to promote the development and use of selective, environmentally safe, and cost-effective fishing gear and techniques. The WCPFC has adopted a number of taxa-specific measures to meet these obligations, as described in the following sections. The Commission administers a regional observer program that collects data on catches of non-target species and on discards. Its scientific data agreement requires that members provide total estimated catches of certain non-target species. The WCPFC maintains a Bycatch Mitigation Information System to facilitate information sharing related to bycatch and bycatch mitigation.

C. Specific Taxa

Sea Turtles

The United States has listed all sea turtles as either threatened or endangered under the ESA. The U.S. Fish and Wildlife Service and NMFS jointly manage sea turtles domestically and collaborate regularly on international conservation activities. The most up-to-date information on the current listing status and designation of critical habitat is available at www.nmfs.noaa.gov/pr/species

Throughout their range, sea turtles are incidentally caught or entangled in fishing gear including pelagic longline, purse seine, trawl, gillnet, pound net, hook and line, and trap/pot fisheries. Sea turtles migrate vast distances over the course of their lives, making them vulnerable to interactions with fishing gear in coastal areas as well as on the high seas.

The United States has several mechanisms to work with countries to mitigate these interactions. For instance, the Shrimp-Turtle law (Section 609 of P.L. 101-162) encourages other nations to take comparable regulatory measures to protect sea turtles in their wild-caught shrimp fisheries, particularly using turtle excluder devices (TEDs). Over the past two decades, the United States has worked with many governments to establish TEDs programs. Each year DOS and NMFS experts carry out TEDs inspections and training across the globe. On May 14, 2014, DOS certified to Congress that the following 14 nations have comparable regulatory programs to reduce sea turtle bycatch to the United States: Colombia, Costa Rica, Ecuador, El Salvador, Gabon, Guatemala, Guyana, Honduras, Mexico, Nicaragua, Nigeria, Pakistan, Panama, and Suriname.

Twenty-six nations and one economy have shrimp fishing environments that do not pose a danger to sea turtles. Of these, 10 nations and one economy harvest shrimp using manual rather than mechanical means, or use other shrimp fishing methods not harmful to sea turtles. They are the Bahamas,

Belize, the Dominican Republic, Fiji, Hong Kong, Jamaica, Oman, the People's Republic of China, Peru, Sri Lanka, and Venezuela. The 16 other nations have shrimp trawl fisheries in cold waters, where the risk of taking sea turtles is negligible: Argentina, Belgium, Canada, Chile, Denmark, Finland, Germany, Iceland, Ireland, the Netherlands, New Zealand, Norway, Russia, Sweden, the United Kingdom, and Uruguay. Specific fisheries in Australia and France have been deemed eligible to export product to the United States. In 2014, Brazil's certification for northern shrimp was withdrawn, making their wild-caught shrimp ineligible to access the U.S. market.

In addition to the Shrimp-Turtle law, the United States has worked aggressively through RFMOs, multilateral environmental agreements, and other forums to urge nations to implement measures comparable to those applicable in the United States to protect sea turtles from fisheries operations. More information about U.S. efforts can be found in the RFMO section below.

Multilateral Sea Turtle Arrangements. With U.S. leadership, two multilateral arrangements are in place to conserve and protect sea turtles. The Inter-American Convention for the Protection and Conservation of Sea Turtles (IAC), as a binding treaty, serves as an important vehicle for countries in the Americas to coordinate their conservation and recovery efforts. Parties are obliged to reduce, to the greatest extent practicable, incidental capture, retention, harm, or mortality of sea turtles in the course of fishing activities. Given the different capacities of parties in the region, the IAC has focused on building national capacities to conserve and recover sea turtle species. For instance, at the Conference of Parties in Galapagos, Ecuador, in June 2013, the parties approved recommendations to improve Guatemala's and Panama's management of exceptions from the Convention's harvest prohibitions. An exception allows for domestic egg harvest for traditional, subsistence communities. The IAC is also coordinating efforts to reduce bycatch of Eastern Pacific leatherback sea turtles in coastal fisheries and to raise awareness about the decline of this species. Finally, the IAC has agreed that all countries will identify index nesting beaches and report data from those beaches, so they can systematically document trends for all sea turtle species in the region.

The Indian Ocean Southeast Asian Marine Turtle MOU, a non-binding instrument, recommends more general conservation action, such as measures to prevent bycatch of sea turtles, but without specifying gear types or actions. The 35 Signatory States meet biennially to evaluate implementation of its Conservation and Management Plan, as well as to identify new actions for the work program. The United States chaired the September 2014 Signatory States meeting in Bonn, Germany. There, the Network for Sites of Importance to Marine Turtles was launched, with 10 sites in nine countries. These critically important sites will now be the focus of conservation and capacity-building efforts. In addition, participants agreed to the 2015-2017 work plan, which includes establishing a working group on IOTC fisheries.

ICCAT's SCRS Subcommittee on Ecosystems met in 2013 to begin work on an ecological risk assessment for sea turtles caught as bycatch in ICCAT fisheries. At the annual meeting in 2013, ICCAT adopted a recommendation that provides greater specificity in safe handling practices (e.g., for the use of line cutters and dehooking devices.) In 2014, several parties provided new productivity information for sea turtles, and the Subcommittee on Ecosystems was made aware of detailed nesting data from the IAC. The Subcommittee agreed to continue to assess the impact of ICCAT fisheries on sea turtles.

Dolphins

The 2013 Report to Congress describes the history and objectives of the AIDCP. Nations and entities that have acceded to or ratified the Agreement include Belize, Colombia, Costa Rica, Ecuador, El Salvador, the EU, Guatemala, Honduras, Mexico, Nicaragua, Panama, Peru, the United States, and Venezuela. Bolivia and Vanuatu apply the Agreement provisionally. The observed dolphin mortality in the EPO purse seine fishery for 2012 was 870; for 2013, it was 801. The AIDCP will examine revised stock mortality limits for common dolphins over the next several years.

Other Marine Mammals

The bycatch of marine mammals in fisheries is a significant factor in long-term conservation and management of marine mammal stocks worldwide. Hundreds of thousands of these animals are killed each year through entanglement in fishing gear. Marine mammals interact with or are bycaught in gillnet, trap, longline, and trawl fisheries. Accurate abundance and bycatch estimates for marine mammals are lacking in areas where marine mammal distribution overlaps with coastal and international fisheries, which makes quantitative analysis of bycatch extremely difficult. Progress in quantifying fishery impacts on marine mammal populations and related efforts to mitigate or reduce mortality have been slow, sporadic, and limited to a few specific fisheries or circumstances.

In 2013, the IOTC adopted a CMM to prohibit vessels from setting purse seines on a school of tuna associated with a cetacean. In the event a cetacean is unintentionally encircled in a purse seine net, the vessel captain must take steps to ensure the cetacean's safe release.

IX. Shark Conservation and Protection

The key components of a comprehensive framework for international shark conservation and management have already been established in global agreements and organizations, which have identified or adopted provisions or guidance to assist States and RFMOs in the development of measures to conserve and sustainably manage sharks. Some of these mechanisms have created international legal obligations with regard to shark conservation and management, while others are voluntary.

A. Global Forums

CITES

Several proposals to list shark species in Appendix II of CITES were considered and adopted at CoP16. These species were proposed for listing because they had experienced significant declines, with international trade considered to be driving their decrease. While some RFMOs have adopted measures to manage sharks, these regional measures alone cannot ensure that the international trade of these species is globally sustainable. Not all range countries are members of RFMOs, while many marine species that are traded internationally are highly migratory, often crossing national boundaries. For these species, conservation can only be achieved by working collaboratively at the global level.

Adoption of species proposals at CITES requires a two-thirds majority of parties present and voting by secret ballot. Leading up to and during CoP16, the United States worked with a coalition of CITES parties committed to gaining support for these proposals, including Brazil, Colombia, Costa Rica, Ecuador, the EU, and Honduras. Strong support from others, particularly countries in West Africa, including Senegal and Sierra Leone, was instrumental to their adoption. The proposals to list sharks in CITES Appendix II were adopted with an annotation to delay entry into effect by 18 months to resolve technical and administrative issues. The effective date for these listings was September 14, 2014. Here are the details on each Appendix II listing proposal:

The United States joined Brazil in co-sponsoring a proposal by Colombia to list the oceanic whitetip shark, a top predator with one of the widest ranges of any shark. In the past, oceanic whitetip sharks were described as among the most common sharks found in temperate, tropical waters; however, populations of this species have declined markedly. The primary threats to oceanic whitetip sharks are unsustainable harvest aimed at supplying the international shark fin market, and mortality from bycatch in other fisheries. At CoP15, in 2010, the United States had submitted a similar proposal; it received a simple majority of votes, but not the needed two-thirds. Since then, a stock assessment reaffirmed that certain populations of this species have continued to decline. The new proposal was adopted at CoP16, with a large number of CITES parties publicly announcing their votes, several citing transparency as their motivation for disclosing their positions.

Brazil, Costa Rica, and Honduras, joined by Colombia, Denmark (on behalf of the EU), Ecuador, and Mexico, proposed listing scalloped, great, and smooth hammerhead sharks on Appendix II. The United States offered a similar proposal at the last CITES meeting due to concerns that over-exploitation for the international fin trade is undermining the conservation status of these species.

The fins of these hammerhead shark species are among the most valuable in trade. That proposal failed to acquire the two-thirds majority needed for adoption. The United States shared concerns about the status of scalloped, great, and smooth hammerhead sharks and supported the leadership of the nations that sponsored the CoP16 proposal for inclusion in Appendix II. The proposal was adopted at CoP16; a large number of CITES Parties again publicly announced their votes.

Denmark (on behalf of the EU), joined by Brazil, Comoros, Croatia, and Egypt, proposed to list the porbeagle shark on Appendix II. This species had also been proposed, unsuccessfully, for listing at the last two CITES meetings. Porbeagle sharks face overexploitation due to demand for their meat. The United States supported this proposal, which was adopted at CoP16.

Convention on the Conservation of Migratory Species of Wild Animals

The United States is not a party to the CMS; however, non-parties are able to participate in individual instruments – MOUs and agreements – concluded under the CMS umbrella. The first meeting of signatories of the CMS Sharks MOU was held in September 2012 in Bonn, Germany. The signatories adopted a conservation plan to catalyze regional initiatives to tackle overfishing and raise awareness of other threats to migratory sharks. The second meeting of signatories will be held in 2015.

B. RFMOs

At its 2013 annual meeting, ICCAT adopted a proposal to provide for the biological sampling of shark species that are prohibited from retention in ICCAT fisheries and that are dead at haul-back, including oceanic whitetip, bigeye thresher, silky and scalloped, smooth, and great hammerhead sharks. ICCAT parties wanting to retain these prohibited shark species for scientific sampling must notify the SCRS and report annually on progress and results achieved through the research.

For the fifth year in a row, the United States co-sponsored a proposal to require that all sharks be landed with their fins naturally attached. At the 2014 annual meeting, co-sponsors of the fins-attached proposal included: Belize, Brazil, Cote d'Ivoire, the EU, Gabon, Ghana, Guatemala, Egypt, Panama, Sao Tome and Principe, Senegal, South Africa, and Trinidad and Tobago. As in past years, no consensus could be reached, but the increasing number of co-sponsors indicates growing support among ICCAT parties for a fins-attached approach.

Also at the 2014 ICCAT annual meeting, the United States introduced a proposal for Atlantic shortfin mako sharks that would have set an annual stock-wide catch limit in both the North and South Atlantic. This proposal was supported by co-sponsorship from the EU, Mauritania, Republic of Guinea, and Senegal, but there was no consensus. Instead, the agreed measure requires parties to improve domestic data reporting systems and provide additional information to ICCAT about how they monitor and manage catches of shortfin mako sharks.

In 2004, NAFO set a 13,500 metric ton total allowable catch limit for thorny skates, a number far in excess of scientific advice for this stock. Although catches remain below the catch limit, the potential for overharvest is considerable. The United States continued to advocate for greater protection for thorny skates. NAFO agreed on lower limits for 2011 and 2012, and has kept decreasing them to 7,000 metric tons for 2013 and 5,000 metric tons for 2014.

The WCPFC has a CMM that includes a prohibition against shark finning, a list of "key shark species," and a research plan for conducting stock assessments for those species. In 2013, the WCPFC adopted a CMM that prohibits retaining on board, transshipping, storing on a fishing vessel, or landing any silky shark caught in the Convention Area, in whole or in part. In addition, the measure requires the release of any silky shark as soon as possible after it is brought alongside the vessel, in a manner that results in as little harm to the shark as possible. The measure is similar to one adopted in 2012 for the oceanic whitetip shark. Although these measures provide some protection, there is growing evidence that additional action is needed to return these stocks to sustainable levels in the western and central Pacific Ocean. In 2014, the Commission adopted further gear restrictions for longline vessels.

While CCAMLR bans directed fishing for sharks, except for scientific research, and requires to the extent possible the live release of incidentally caught sharks, shark finning is not prohibited. At the 2013 and 2014 meetings, the United States proposed requiring any retained sharks to be landed with all fins naturally attached, and other provisions to encourage collection of information about sharks and shark habitat. While many members supported the proposal, Japan and China blocked its adoption.

In 2013, the IATTC adopted a measure to prohibit vessels from setting purse seines on a school of tuna associated with a live whale shark. Similar to the bans on encirclements of cetaceans, the IATTC whale shark measure requires captains to ensure safe release in cases where whale sharks are not deliberately encircled. At the 2014 annual meeting, the EU sponsored a proposal to require sharks to be landed with their fins naturally attached. Despite arguments from the United States and others that the 5 percent fin-to-carcass ratio in the existing resolution is not adequate to prevent shark finning, Japan and China disagreed and blocked consensus on the proposal. The EU also sponsored a proposal to prohibit retention of silky sharks in IATTC fisheries, similar to the WCPFC measure adopted in 2013. The United States provided comments on the proposal and supported the measure along with several other countries. Costa Rica, Ecuador, and Guatemala, however, requested an exemption for artisanal longline vessels under 24 meters. The EU argued that such an exemption would make the measure ineffective and withdrew the proposal.

X. International Cooperation and Assistance

The international community recognizes the importance of providing necessary tools and training to assist developing coastal and fishing States with management and monitoring of their fisheries and fishing vessels. Such assistance helps nations address IUU fishing activities, promotes the adoption of measures to mitigate the adverse impacts of fishing activities on PLMRs, and furthers shark conservation programs. The need for such cooperation and assistance is recognized in global and regional fisheries instruments, including the UNFSA.

A. International Institutional Efforts

ICCAT Funds

ICCAT has several funds created specifically for scientific capacity building; these are used primarily to finance travel of scientists from developing States to participate in intersessional scientific meetings and the annual SCRS meeting. In 2013, ICCAT established a mechanism for scientists from developing States to attend ad hoc training for up to 14 days. A separate fund supports the attendance of developing State members at meetings of the Commission. ICCAT adopted a U.S. proposal in 2014 to create a fund to assist developing States in carrying out their responsibilities under ICCAT's Port Inspection Scheme.

CCAMLR Activities

Researchers from Poland and Russia received scholarships in 2013 and 2014, respectively, under the CCAMLR Scientific Scholarship Scheme. The CCAMLR Secretariat and Australia provided capacity-building training to representatives from the Hong Kong Special Administrative Region on the use of the electronic CDS and relevant conservation measures.

WCPFC Efforts

The WCPFC is the only RFMO whose budget contains a line item, funded by all members, to support the special needs of developing States parties.

IATTC Fund

In 2011, the IATTC created a fund for strengthening the scientific and technical capacity of developing countries, which will allow them to fully comply with their obligations under the Antigua Convention. In 2014, the IATTC amended this measure to require an annual contribution of 2 percent of its budget to this fund.

B. Bilateral and Regional Assistance

Through the MSA and other authorities, NMFS engages in international cooperation and assistance, with particular emphasis on efforts to combat IUU fishing, mitigate bycatch of PLMRs, and conserve sharks. The United States shares stocks of living marine resources, including protected resources, with other countries. Many living marine resources cross national maritime boundaries and venture into the high seas during their lives. Conservation activities or the lack thereof in countries other than the United States can either enhance or undermine our own conservation efforts. Management measures in other countries can directly affect the status of fish resources the United States harvests. In addition, lack of conservation efforts can interfere with the recovery of endangered or threatened species wherever they occur. Because the U.S. seafood market is increasingly reliant on foreign supplies of fish and fish products, it is in the U.S. interest to promote the sound conservation and management of all fisheries and aquaculture operations that supply the U.S. market.

The United States is a member of many global and regional marine conservation organizations. Decisions on management measures are made either by consensus or require a majority of the countries present to support the measure. By contributing to the capacity of member countries to manage marine resources sustainably, we increase the number of countries that are willing to adopt and implement management measures to accomplish these goals.

More fundamentally, the dependence of the U.S. market on imports of wild-harvested and farmed seafood, and the growing demands of American consumers for assurance that fish are not the product of illegal or unsustainable practices, require the United States to address the lack of fisheries management and enforcement capacity in many developing countries. NMFS' assistance efforts strengthen international fishery management organizations; build strategic partnerships with other agencies, nations, and donors; and enable other nations to become better stewards.

The program established by the NMFS Office of International Affairs and Seafood Inspection focuses on priority geographic areas and important programmatic issues. In establishing the program, NMFS recognized that it should focus on particular regions of the world to take advantage of economies of scale where other work was already being done and to take advantage of funding made available from other sources. In West Africa, one of the areas where IUU fishing is most prevalent, funding and assets in the region were provided by the U.S. Africa Command. In the Coral Triangle Region (Indonesia, Malaysia, Papua New Guinea, the Philippines, the Solomon Islands, and Timor Leste), the United States Agency for International Development (USAID) provided funding and sought NOAA as a partner for this work. As our nearest neighbors, Latin America and the Caribbean share with the United States many marine species, both targeted and taken as bycatch. Funding and other support for work in Central America is available from USAID, DOS (for work under the auspices of the Central America-Dominican Republic Free Trade Agreement), and from the U.S. Southern Command for work in Colombia.

In addition to the regional focus, capacity-building activities are undertaken within specific program areas, of which the chief ones are marine mammal, sea turtle, and shark conservation; CITES-listed marine species; and the IWC action plans.

This section sets forth some examples from among the many programs NMFS carried out during 2013 and 2014.

Strategies to Combat Illegal Fishing in Central America

Since 2008, NOAA and the USCG have worked to strengthen fisheries MCS in Colombia. In April 2014, a workshop in Cartagena brought together military and civilian authorities to consider ways to combat IUU fishing. Funded by Conservation International, NMFS, and USAID, the workshop was convened by the Colombian Ocean Commission. About 80 Colombian officials attended, including from the fisheries authority, the Ministry of Environment, the National Park Service, and the Coast Guard.

Representatives of NOAA, INTERPOL, and the USCG made presentations and participated in working groups to share best practices and consider future efforts.

In December 2014, NOAA participated in an INTERPOL-sponsored fisheries workshop in Panama. The conference included participants from Colombia, Costa Rica, the Dominican Republic, Ecuador, El Salvador, Guatemala, Nicaragua, Panama, and Peru. The conference addressed tools and methods available to identify IUU fishing activities in international waters off Central America, and INTERPOL's role in providing assistance. NOAA trained attendees on evidence collection and documentation, decision making, case preparation, and developing and prioritizing goals relevant to fisheries enforcement operations. NOAA also provided guidance on inspections of commercial fishing vessels, complex investigations, and town hall meetings with the commercial fishing community.

Regional Collaboration on the Conservation of Queen Conch

International trade in queen conch is regulated under Appendix II of CITES. NMFS, by sponsoring workshops, has encouraged countries such as the Bahamas, Belize, and Colombia to promote coordination between CITES and regional fisheries authorities, and to further cooperation among range States in enforcement of national and CITES requirements. A working group consisting of the Caribbean Fishery Management Council, OSPESCA (the regional fisheries organization in Central America), the Caribbean Regional Fisheries Mechanism, and WECAFC made recommendations for the sustainable and legal management of this species. Several complementary decisions on queen conch conservation were adopted at CoP16. At its 2014 meeting, WECAFC passed a resolution requesting that its members prepare and share their CITES non-detriment findings and committing to develop a regional plan for management and conservation.

CITES Workshops

In fulfillment of commitments made at CoP16, NMFS provided support for several workshops to assist CITES parties with implementation of the shark listings that became effective in September 2014. These workshops brought together CITES and fisheries authorities to promote interagency collaboration and exchange of information.

To assist Latin American and Caribbean countries with these new international requirements, the Government of Brazil hosted a regional workshop in December 2013. Experts covered a range of topics at the workshop based on needs identified by the countries at the last CITES meeting, including training on the use of tools and techniques for the identification of sharks and shark products in trade, demonstration of monitoring products from the point of harvest to the point of export through real-life examples, and international shark conservation and management measures. The workshop identified regional needs and priorities for Latin American and Caribbean countries for implementing the CITES shark listings to help guide future activities.

In August 2014, the Government of Senegal organized a regional workshop, in collaboration with the Government of Sierra Leone, with the support of NMFS and the Sub-Regional Fisheries Commission. At the meeting, participants from 13 West African countries shared information and received identification training and tools that will facilitate implementation of the CITES shark and ray listings. The workshop participants adopted an action plan identifying recommendations to address the priority needs of the region for the implementation and enforcement of the shark and ray species listed in CITES. This workshop provided a foundation for future collaboration among West African countries to help ensure that trade of shark species in this region is sustainable and does not threaten their survival.

The Government of Colombia also hosted an international CITES workshop in November 2014, which focused on regulating the trade of CITES-listed shark products and strategies for making a sustainability finding required under CITES. More than 60 participants representing some 20 countries discussed mechanisms for monitoring and controlling the trade of CITES-listed shark species and the process for making Non-Detriment Findings within the context of shared international experiences and case studies from different parts of the world. Subsequently, the participants identified recommended future actions needed to effectively implement the recent shark listings.

Cooperation with Mexico on Law Enforcement

As a result of the 2013 bilateral fisheries meeting between the United States and Mexico, NOAA hosted a fisheries law enforcement meeting in Houston in July 2014. Participants discussed ways to collaborate on addressing IUU fishing, combating trafficking in illegally taken species, and controlling bycatch (particularly of vaquita). This meeting led to coordinated law enforcement operations by Mexico and the United States in September 2014.

Enhancing Fisheries Enforcement, Legal Frameworks, and Observer Capabilities in West Africa

During 2013 and 2014, NOAA has collaborated in efforts to train West African fisheries management and enforcement officials and students from Gabon, Liberia, and the seven member States of the Sub-Regional Fisheries Commission: Cape Verde, the Gambia, Guinea, Guinea Bissau, Mauritania, Senegal, and Sierra Leone.

NOAA conducted an intensive three-week course in February 2013 in Monrovia, Liberia, in partnership with the Liberia Bureau of National Fisheries and the World Bank's West Africa Regional Fisheries Project. The training for 16 observers covered near-shore demersal trawl, tuna purse seine, and longline fisheries. Topics included regulations governing observers; gear, catch/effort, and sampling in the various fisheries; fish, marine mammal, and sea turtle identification and data collection; safety at sea, in the classroom and hands on; and communication and vessel electronics.

NOAA presented a fisheries enforcement and prosecution training in Monrovia, Liberia, in June 2013, for 23 officials from a variety of institutions, including the Liberian Coast Guard and the Ministries of Justice, Agriculture, and Defense. Topics included fisheries management and MCS activities, domestic fisheries legislation, investigation coordination, types of evidence, chain of custody and use of evidence, prosecutorial considerations, recordkeeping and reporting, witnesses and use of observers as witnesses, penalty policy, and settlement negotiations. One day was allocated to table-top exercises, giving participants a chance to apply their new knowledge to practical scenarios.

NOAA provided similar training for 25 officials in Freetown, Sierra Leone, in July 2013, in collaboration with the World Bank and the Sierra Leone Ministry of Fisheries and Marine Resources. Trainees came from five agencies, including the Sierra Leonean Armed Forces, the Ministry of Justice, and the Sierra Leonean Police. As with the Liberian training, the objective was to strengthen fisheries enforcement and prosecution capacity and to promote interagency cooperation.

Based on ongoing collaboration with the Government of Liberia and the World Bank, NOAA conducted a training course on case package preparation, better equipping fisheries enforcement inspectors to perform their key role in prosecuting fisheries violations in national waters. The workshop in July 2014 included fisheries enforcement agents and personnel from relevant Liberian government agencies. The training addressed all aspects of a fisheries investigation: monitoring activities, identifying violations, collecting information, and building a final investigative case package for prosecutors.

In 2014, NOAA worked in close consultation with the Sub-Regional Fisheries Commission staff toward presenting training in September in Dakar, Senegal, to improve the capacity of the Commission to advise its member States on legal issues in fisheries management. The objective of the project is to provide Commission staff, as well as legal advisers and fisheries managers from the member States, a better understanding of the major international instruments, binding and non-binding, relating to management of ocean fisheries; proper implementation and enforcement of domestic fisheries legislation, consistent with international law; and mechanisms of cooperation between flag, port, and coastal States in evidence collection, identification of violations, and referrals for legal proceedings. The Ebola crisis in the region necessitated postponement of the training until 2015.

In 2014 NMFS conducted research on TEDs in Gabon as a means of evaluating TED performance and expanding turtle bycatch mitigation methods in Gabonese trawl fisheries. NMFS also provided assistance for Gabon to host a sub-regional TED training workshop for the region, including Cameroon, Equatorial Guinea, Ghana, and Nigeria. As Gabon has been operating a TED project for several years, the intent of the workshop was to allow Gabon to showcase its TED research results and the administrative strategy that contributed to a successful TED program. This workshop has also been postponed until 2015, due to the Ebola crisis in the region.

In 2014, NMFS worked with the Sub-Regional Fisheries Commission to enhance the members' shark conservation and management efforts. The goal of the project is to update and further the implementation of national plans of action for sharks in Cape Verde, The Gambia, Guinea, Guinea Bissau, Mauritania, Senegal, and Sierra Leone, and to promote a regionalized fins-attached policy. The Commission made site visits to all but one member country to discuss standardized shark data collection and a strategy for fins-attached policies. A regional workshop will be held some time in 2015.

Asia Pacific Capacity Building

The interrelated themes of the Asia Pacific capacity-building portfolio are mitigating the negative impacts of the Asian tropical trawl fishery and promoting the ecosystem approach to fisheries management. Balancing regional efforts with specific projects in individual countries allows for the greatest impact. Pilot projects on the national level attract the attention of the FAO, which can then expedite similar capacity-building programs in the region. Many of these projects also support the USAID/NOAA-funded Coral Triangle Initiative (2009–2014), which includes Indonesia, Malaysia, Papua New Guinea, the Philippines, Solomon Islands, and Timor Leste.

To address Asian trawl fisheries, NOAA supported an FAO/Asia-Pacific Fishery Commission expert workshop in Thailand in 2013. The workshop developed practical, easy-to-read guidelines for managers of Asian tropical trawl fisheries, one of the most productive fisheries in the world. Today the trawl sector throughout the region catches enormous quantities and diversities of fish and shrimp to fulfill market demands, with very little discarding. The guidelines, if widely followed, will mitigate overcapacity, conflict between commercial and artisanal fisheries, excess fishing effort, and (because trawls are poorly selective) the sub-optimal harvest of juveniles of many species. The final version of guidelines was published in June 2014; see www.pifsc.noaa.gov/cred/eafm_training

In 2014, NOAA awarded a grant to the University of Washington to collect background on trawl fishery legislation in three countries in South and

South East Asia, through interviews with stakeholders. The goal is to develop measurable social and ecological performance indicators.

To advance the ecosystem approach to fisheries management, NOAA, FAO, and the Bay of Bengal Large Marine Ecosystem Project developed a course that was presented in Sabah, Malaysia, in 2013. The pilot presentation tested the training material and obtained feedback for development of a regional pool of trainers in the approach.

Through a partnership with the World Bank and Vietnam's Ministry of Agriculture and Rural Development, NOAA is helping Vietnam build capacity to manage its coastal resources. A NOAA team ran week-long exercises in the cities of Vinh (2013), which focused on the development of a management framework, and Nha Trang (2014), which specifically looked at stakeholder identification, prioritization, and engagement.

A NMFS employee and the chief scientist of the Mid-Atlantic Fishery Management Council traveled in February 2014 to Indonesia, to present non-central models of management to stakeholders including central, provincial, and district officials. The group focused on basic management of the blue swimming crab fishery, which is Indonesia's third largest fishery by export value, at $350 million per year. NMFS led a second exchange of experts on small-scale fisheries management in Semarang, Indonesia, in 2014. All the stakeholders in the fishery developed a comprehensive management framework that accounted for the local realities of newly decentralized fisheries management powers, inexperienced district fisheries officials, and other stakeholders who had not previously engaged one another.

The outcome was that fishermen from the community already targeted by this project received a legal decree from the district government endorsing the NOAA-facilitated template as a planning tool. The second community, which NOAA invited because it had not yet benefited from outreach, made a similar request for endorsement to its local government. Through this peer-topeer approach, NMFS hopes to expedite the use of basic management planning principles throughout coastal communities. The last part of the workshop was a half-day presentation to more than 150 University of Diponegoro students, given in conjunction with the fisheries representative from the Japanese International Cooperation Agency.

Promoting Observer and Monitoring Programs

The International Fisheries Observer and Monitoring Conference provides a forum for observers, researchers, managers, and industry to collaborate on future directions for fisheries observers and monitoring systems. Topics at the conference held in Viña del Mar, Chile, in April 2013, included the role of electronic monitoring programs and the challenges of observing in artisanal fisheries. NOAA staff served on the steering committee and participated in several sessions.

In December 2013, NOAA conducted a workshop in Korea, as part of the NOAA-Korea Joint Project Agreement. The goal of the project is to develop and implement a fisheries observer program for the Korean purse seine mackerel fishery.

Reducing Marine Mammal Bycatch

During this reporting period, NMFS has worked with international partners to reduce bycatch of marine mammals through gear modification and other capacity-building activities, as detailed below.

NMFS supported a team of experts, the International Committee for the Recovery of the Vaquita, that has been assisting Mexico in efforts to implement a recovery plan for the critically depleted vaquita, including estimating vaquita abundance, evaluating the potential of proposed vaquita protected areas in the northern Gulf of California, providing expert advice on the development of vaquita acoustic monitoring programs, and conducting research to develop vaquita-safe gear. In August 2014, the Committee, reporting that only 97 vaquita remain, recommended an immediate ban on gillnetting throughout the entire range of the vaquita.

On December 23, 2014, Mexican authorities proposed a $37 million plan to ban gillnet fishing in most of the upper Sea of Cortez. The new proposal would substantially increase the area closed to fishing to include most of the vaquita's habitat. If the two-year plan is put into place after public review, some of the $37 million would be used to compensate fishermen affected by the ban, including paying some of them to patrol the closed area looking for violations. Because Mexico has failed to effectively enforce its regulations prohibiting fishing within the much smaller refuge established in 2009, increased enforcement efforts would be critical to the success of the proposed plan.

In August 2013, the Executive Secretariat of the Action Plan for the South East Pacific, with the cooperation of NMFS, convened a workshop in Salinas, Ecuador. Humpback whales in coastal areas of Colombia, Ecuador, Panama, and Peru, as well as blue whales along the coasts of Chile and Peru, are vulnerable to ship strikes and interaction with fishing gear. The main objectives of the workshop were to identify areas of whale concentration and conduct risk analysis.

NMFS and the New England Aquarium's Consortium for Wildlife Bycatch Reduction convened a workshop in October 2013 to review research on reducing bycatch and subsequent mortality of odontocetes (toothed whales) in longline fishing operations worldwide. Participants discussed measures such as time-area closures, acoustic deterrents, weak hooks, and net sleeves. The project will produce best practices guidelines to reduce marine mammal bycatch in longline fisheries and an on-line clearinghouse of information on research carried out to date. As many as 16 original research papers drawn from presentations made at the workshop, including the summary report, will be included in a special issue of the ICES Journal of Marine Science.

In December 2013, NMFS hosted a workshop to address bycatch in Pacific fisheries, at a meeting of the Society of Marine Mammalogy in Dunedin, New Zealand. Workshop participants identified key informational needs and initial actions required to reduce marine mammal bycatch, and identified regions, fisheries, or situations that are ripe for testing mitigation strategies or developing pilot programs to mitigate marine mammal bycatch.

NMFS supported, through the Peruvian non-profit group ProDelphinus, capacity-building efforts in 2014 to raise awareness among small-scale driftnet fishermen of the need to reduce incidental captures of small cetaceans, and to purchase and distribute acoustic alarms (pingers) for this purpose. The project held 13 workshops, presenting information to 102 attendees in 10 ports. Through this grant ProDelphinus distributed 101 pingers to 10 fishing vessels. In return for the pingers, fishermen provide reports on their fishing activity and associated bycatch.

Marine Mammal Conservation

NMFS has also supported other capacity-building projects relating to marine mammals, as set forth below.

NMFS published a report in 2014 summarizing 10 years of research and conservation for endangered southern resident killer whales. Working with partners at the Whale Museum, the transboundary Pacific Whale Watch Association, and Canada's Department of Fisheries and Oceans, NMFS will repurpose the report into a resource for whale-watch naturalists, who are in a unique position to educate passengers about simple actions they can take to support recovery. This project builds transboundary capacity with Canada and helps implement both the recovery plan for southern resident killer whales and the International Marine Mammal Action Plan.

In 2014, NMFS launched an effort to collaborate with Mediterranean monk seal researchers and managers, to share expertise on this critically endangered species. In particular, NMFS will share its robust scientific and veterinary expertise to help the European managers better monitor the population, determine threats, and refine recovery strategies through a two-week meeting in April/May 2015.

NMFS, in partnership with the Woods Hole Oceanographic Institution and the International Fund for Animal Welfare, organized an "International Marine Mammal Stranding and Entanglement Response Toolkit Development Workshop" at Woods Hole in June 2014. Key marine mammal experts from Central and South America, Europe, South Asia, and the United States identified existing training resources for consolidation into a web-based platform, defined and developed core skills and training objectives for different audience categories, and discussed effective approaches to develop core skills and implement best practices.

In 2014, NMFS supported the New England Aquarium's Consortium for Wildlife Bycatch Reduction in its research on pingers, which are used in U.S. gillnet fisheries to reduce bycatch of harbor porpoise, common dolphins, and beaked whales. At present, pingers are the only tools available to gillnet fishermen for reducing marine mammal bycatch. This project is evaluating application of pingers in fisheries around the world. In small-scale non-industrial fisheries, the relatively high expense of purchasing and maintaining pingers is an obstacle to their use. In Argentina, where previous research showed that pingers deter endangered Franciscana dolphins from becoming entangled in gillnets, investigators are researching whether fewer pinger units can still achieve a bycatch reduction effect. Experimental pingers have been acquired from overseas, fishermen identified for carrying out the experiment, and a team of observers lined up for the field trial that began in September 2014. Also, a trial is planned for 2015 for one of three candidate species of cetaceans (Irrawaddy dolphin, finless porpoise, or Dall's porpoise) to see whether pingers create an area displacement effect, an indication that they might work as a bycatch deterrent.

Sea Turtle Conservation

NMFS research has focused on the development and use of bycatch reduction technologies for gillnet fisheries and trawl fisheries. For gillnets, net illumination has been shown to reduce sea turtle interactions while maintaining target catch rates and catch value. For trawl fisheries, TEDs have been shown to be an effective measure. Escape devices for pound net fisheries are also under investigation. Through 2014, sea turtle conservation projects continued in Chile, El Salvador, Mexico, Japan, Nicaragua, and Peru, and expanded to Indonesia and Malaysia, where they provided opportunities to collaborate and build in-country capacities of local government fishery managers, academics, conservation NGOs, and local fishing cooperatives.

NMFS has worked extensively in Baja California, Mexico, developing a variety of bycatch reduction technologies for small-scale gillnet fisheries. Previous work has shown that illuminated nets can reduce green sea turtle interaction rates from 40 to 60 percent using ultraviolet (short-wavelength) and green (mid-wavelength) with no adverse effect on target fish catch rates or fishermen profits. Through 2013 and 2014, studies have extended this work into the longer wavelengths and shown that catch composition can be influenced by the wavelength of light used. Such illumination strategies can also reduce the bycatch of some shark species.

NMFS scientists, in collaboration with a Mexican small-scale fishing community, are examining the effectiveness of using inexpensive video technology to conduct electronic monitoring in small-scale gillnet fisheries in Mexico. In 2014, a scoping study compared the effectiveness and cost of collecting catch data from onboard observers and from video systems. Results indicate similar empirical values for retained and discarded species, suggesting that these cameras may be a reliable substitute for observers in these fisheries, provided that personnel are available to analyze the video data. If managers need additional information, for example from biopsies, fisheries observers would be required.

Throughout 2013–2014, NMFS collaborated with ProDelphinus and local fishing communities in Peru to test sea turtle bycatch mitigation strategies in gillnet fisheries, which have a very high bycatch rate of sea turtles. Results from the study show that net illumination in the gillnet fishery of northern Peru reduces sea turtle bycatch rates by more than 50 percent, with no change to target fish catch rates. In addition, results also indicate that this technology can be used to reduce seabird bycatch in the same gillnet fleet. Beginning in 2014, testing this net illumination has been expanded into driftnet fisheries. These driftnet fisheries target swordfish and shark species, but interact with leatherback, loggerhead, and other sea turtle species as well.

Hawksbill turtles inhabiting the EPO are among the most endangered sea turtle populations. Small-scale lobster gillnet fisheries have been identified as a major threat to hawksbills in the eastern Pacific, in particular those based in El Salvador and Nicaragua. The use of light-emitting diodes to illuminate nets has proven an effective tool to reduce marine turtle bycatch. Beginning in 2014, NMFS worked El Salvadorian and Nicaraguan NGOs to begin bycatch reduction trials.

Beginning in 2013, NMFS collaborated with NGOs, government agencies, and academics in Indonesia to better understand the impacts of the region's small-scale fisheries on protected marine megafauna. An assessment of the gillnet fisheries based in Paloh, West Kalimantan (Borneo) suggested that several hundred sea turtles, including green, hawksbill, leatherback, and olive ridley sea turtles, are caught each year in those fisheries. NOAA supported a workshop in 2013 that concluded with a plan to conduct bycatch mitigation experiments in the sea turtle "hot spot" off Paloh. In early 2014, NMFS researchers conducted a training workshop in methods to characterize the gillnet fisheries, collect catch data, disentangle and measure sea turtles, tag sea turtles, and conduct experimental trials.

Since 2008, NOAA has collaborated with the Sea Turtle Association of Japan to develop, design, and support gear mitigation research to reduce bycatch of loggerhead turtles in coastal pound net fisheries in Japan. During 2014, pound net escape device trials continued at the Suma Aquarium, but also moved into the field to test gear mitigation measures in an operating fishery. This first in-situ trial assessed how the device might operate in an operating pound net, collected environmental and technical data regarding its functionality, collected data regarding retention of target species, and tested the data recording and camera equipment.

ANNEX 1: INTERNATIONAL FISHERIES AND RELATED AGREEMENTS AND ORGANIZATIONS TO WHICH THE UNITED STATES IS A PARTY OR IN WHICH THE UNITED STATES HAS A SUBSTANTIAL INTEREST

To provide basic knowledge of the multilateral agreements, RFMOs, and related international organizations concerning living marine resources of which the United States is a member or that are of substantial interest to the United States, a list of many such organizations and agreements, with brief descriptions, is set forth below.

Global

United Nations Convention on the Law of the Sea. This treaty sets the rules for jurisdiction and management authority in the oceans, and establishes general requirements concerning conservation. The Convention currently has 166 parties; the United States is not yet a party, but operates consistent with the fisheries provisions of the Convention. President Clinton submitted the Convention to the Senate in 1994.

Agreement for the Implementation of the Provisions of the United Nations Convention on the Law of the Sea Relating to the Conservation and Management of Straddling Fish Stocks and Highly Migratory Fish Stocks (UN Fish Stocks Agreement or UNFSA). This agreement provides more specific rules for the conservation and management of straddling and highly migratory fish stocks, including application of the precautionary approach, ecosystem-based management, a requirement that nations with vessels fishing on the high seas either join the appropriate RFMO or apply the CMMs established by that RFMO to its fishing vessels, and other similar requirements. The 1995 agreement, which entered into force in 2001, now has 82 parties, including the United States.

Agreement to Promote Compliance with International Conservation and Management Measures by Fishing Vessels on the High Seas (FAO Compliance Agreement). This agreement requires flag States to exercise control over their vessels on the high seas to ensure that they follow applicable conservation and management regulations. The agreement was adopted in 1993 and entered into force in 2003. It has 40 parties, including the United States.

FAO Code of Conduct for Responsible Fisheries. This non-binding document, prepared in 1995, sets forth principles and international standards of behavior for responsible fisheries practices, to ensure effective conservation, management, and development of living aquatic resources.

International Whaling Commission. The IWC was established under the International Convention for the Regulation of Whaling in 1946. The primary function of the IWC is to establish and revise measures that govern the conduct of whaling throughout the world. The Commission currently has 88 parties, including the United States.

Convention on International Trade in Endangered Species of Wild Fauna and Flora. CITES provides for the protection and regulation of certain species of wild fauna and flora, including certain living marine species, against over-exploitation, through limitations on international trade. Under CITES, species are listed in Appendices according to their conservation status:

Appendix I ("threatened with extinction"); Appendix II (may become threatened with extinction unless trade is strictly regulated); and Appendix III (species that any party identifies as being subject to regulation within its jurisdiction for the purpose of preventing or restricting exploitation, and that needs the cooperation of other parties in the control of trade). CITES currently has 180 parties, including the United States.

Agreement on the Conservation of Albatrosses and Petrels. ACAP, a legally binding agreement, was established under the CMS; it has 13 parties. Its purpose is to enhance the understanding of the conservation status of albatrosses and petrels and their susceptibility to a range of threats, as well as to provide an effective means of mitigating those threats. Although not a party, the United States participates in ACAP meetings as an observer.

Memorandum of Understanding on the Conservation of Migratory Sharks. This non-binding agreement, negotiated under the auspices of the CMS, provides an international framework for coordinating sustainable management and conservation efforts for seven species of migratory sharks. The MOU has 38 signatories, including the United States, and two cooperating partners.

Atlantic

International Commission for the Conservation of Atlantic Tunas. ICCAT provides for international cooperation in conservation and management, including scientific research, for tunas and tuna-like species in the Atlantic. It covers all waters of the Atlantic Ocean, including the adjacent seas. ICCAT has 49 contracting parties, including the United States, plus five cooperating non-members.

North Atlantic Salmon Conservation Organization. NASCO has jurisdiction over salmon stocks that migrate beyond areas of coastal State jurisdiction in the Atlantic Ocean north of 36° N throughout their migratory range. It has six parties, including the United States.

Northwest Atlantic Fisheries Organization. NAFO's Convention Area is located within the waters of the Northwest Atlantic Ocean roughly north of 35° N and west of 42° W. The principal species managed are cod, flounders, redfish, American plaice, Greenland halibut (turbot), capelin, shrimp, hake, and squid. NAFO has 12 contracting parties, including the United States.

Southeast Atlantic Fisheries Commission. The SEAFO Convention, which entered into force in 2003, regulates fisheries outside EEZs in the Southeast Atlantic Ocean. Species covered include fish, mollusks, crustaceans, and other sedentary species, except species subject to coastal State jurisdiction and highly migratory species. There are currently seven parties. The United

States signed the Convention, but is not a party because no U.S. vessels fish in the area.

Pacific

Western and Central Pacific Fisheries Commission. The WCPFC manages tuna and other highly migratory species in the western and central Pacific Ocean. The Convention entered into force in 2004. It currently has 26 members, including the United States; seven participating territories; and eight cooperating non-members.

South Pacific Regional Fisheries Management Organization. The Convention on the Conservation and Management of High Seas Fishery Resources in the South Pacific Ocean entered into force on August 25, 2012. Its objective is to ensure the long-term conservation and sustainable use of fishery resources and to safeguard the marine ecosystems in which these resources occur. The Convention has 13 parties; the United States, along with five other nations, is a cooperating non-contracting party.

South Pacific Tuna Treaty. This agreement provides U.S. tuna purse seine vessels access to fish in the waters of the Pacific Island parties to the Treaty, including adjacent high seas areas in the central and western Pacific. Although not a fisheries management arrangement, it is referenced in this report because it contains some important and forward-looking monitoring and control provisions, including observer and VMS requirements. The Treaty has 17 parties, including the United States. It is administered by the Forum Fisheries Agency, comprised of the 16 Pacific Island parties. The treaty has been extended to December 31, 2015.

Inter-American Tropical Tuna Commission. The IATTC manages tunas, tuna-like species, and other species taken by tuna-fishing vessels in the EPO. The Commission has 21 members, including the United States, plus four cooperating non-members.

Agreement on the International Dolphin Conservation Program. This agreement establishes legally binding mechanisms to reduce incidental dolphin mortality in the tuna purse seine fishery in the EPO to levels approaching zero. The agreement has 14 parties, including the United States, plus two nations that apply the Agreement provisionally.

North Pacific Anadromous Fish Commission. The NPAFC promotes the conservation of anadromous stocks (salmon) and ecologically related species, including marine mammals, seabirds, and non-anadromous fish, on the high seas of the North Pacific, the Bering Sea, and the Sea of Okhotsk, north of 33° N. It has five parties, including the United States.

Convention on the Conservation and Management of Pollock Resources in the Central Bering Sea. This Convention was established to conserve and manage the pollock resources in the high seas area of the Bering Sea (the "donut hole"). It has six parties, including the United States.

Pacific Salmon Commission. The PSC implements the United States-Canada Pacific Salmon Treaty. Four commissioners and four alternates from each nation represent the interests of commercial and recreational fisheries as well as federal, state, and tribal governments. The PSC provides regulatory advice and recommendations to the two parties with regard to salmon originating in waters of one country that are subject to interception by the other, salmon that affect the management of the other country's salmon, and salmon that biologically affect the stocks of the other country.

International Pacific Halibut Commission. Established by a 1923 Convention between the United States and Canada, the Commission's mandate covers research on and management of the stocks of Pacific halibut within Convention waters of both countries. The Commission consists of three government-appointed commissioners for each country.

Memorandum of Understanding for the Conservation of Cetaceans and their Habitats in the Pacific Islands Region. Negotiated under the auspices of the CMS, this non-binding MOU provides an international framework for coordinated conservation efforts for cetaceans and their habitats in the Pacific Islands Region. The MOU has 15 signatories, including the United States.

Southern Ocean

Commission for the Conservation of Antarctic Marine Living Resources. With the exceptions of any commercial seal hunt south of 60° S and all whaling activities, CCAMLR conserves and manages all marine living resources between the edge of the Antarctic continent and the Antarctic Polar Front (varying between 45° S and 60° S). There are 25 members of the Commission, including the United States. Another 11 countries have acceded to the Convention. They have agreed to be legally bound by its terms, but do not contribute to the budget or participate in decisions.

Convention for the Conservation of Antarctic Seals. The Convention is designed to promote and achieve the protection, scientific study, and rational use of Antarctic seals, and to maintain a satisfactory balance within the ecological system of Antarctica. It prohibits the killing or capture of seals in the area south of 60° S, except as specifically provided for in the Convention. It has 17 parties, including the United States.

Western Hemisphere

Inter-American Convention for the Protection and Conservation of Sea Turtles. The IAC is the only binding Convention for the protection and conservation of sea turtles in the world. The IAC specifically protects six of the seven species of sea turtles: loggerhead, green, leatherback, hawksbill, olive ridley, and Kemp's ridley. This Convention entered into force in 2001 and has 15 parties, including the United States.

Indian Ocean

Indian Ocean–South East Asian Marine Turtle Memorandum of Understanding. This MOU operates as a non-binding agreement under the CMS. It provides a framework within which the States of the region as well as other concerned States can work together to conserve and replenish depleted marine turtle populations for which they share responsibility. The MOU has 35 signatories, including the United States.

Annex 2: United States Laws Addressing IUU Fishing, PLMR Bycatch, and Shark Conservation, including Summaries of Recent Enforcement Cases

Magnuson-Stevens Fishery Conservation and Management Reauthorization Act of 2006. The 2006 reauthorization of the Magnuson-Stevens Fishery Conservation and Management Act, 16 U.S.C. 1801 *et seq*., directs substantial attention to fishing issues outside U.S. waters, particularly IUU fishing and bycatch of PLMRs. Title IV of the Act amended the High Seas Driftnet Fisheries Moratorium Protection Act, 16 U.S.C. 1826d-1826g, to call on the Secretary of Commerce to urge other nations and RFMOs to address IUU fishing and to put into place regulatory measures to end or reduce bycatch of PLMRs comparable to those of the United States, taking into account different conditions. It also puts into place an identification and certification procedure for nations whose vessels engage in IUU fishing or bycatch of PLMRs.

Magnuson-Stevens Fishery Conservation and Management Act. The MSA, originally enacted in 1976, is the foundational legislation for the conservation and management of fisheries within the U.S. EEZ. Besides establishing the framework for regulating U.S. fisheries, the Act contains specific and extensive prohibitions and enforcement authorities to ensure a high rate of compliance with regulations governing both domestic and foreign fishing within the EEZ.

High Seas Driftnet Fishing Moratorium Protection Act. This Act prohibits the United States from entering into international agreements that would prevent full implementation of the UN Moratorium on Large-Scale High Seas Driftnets. The MSRA and SCA added specific authorities and responsibilities to assist in reducing or eliminating IUU fishing, bycatch of PLMRs, and certain shark fishing practices.

High Seas Driftnet Fisheries Enforcement Act. This Act, 16 U.S.C. 1826a-1826c, seeks to end the use of large-scale driftnets by foreign fisheries operating beyond the EEZ of any nation. Among other provisions, the Act authorizes identification of nations whose vessels are engaging in high seas fishing with large-scale driftnets; such identification may lead to limitations on port entry and on the importation of certain products from those nations.

High Seas Fishing Compliance Act. This Act, 16 U.S.C. 5501-5509, implements the FAO Compliance Agreement for vessels flagged in the United States. The HSFCA requires all U.S. vessels to obtain a permit before engaging in operations on the high seas; authorizes the Secretary of Commerce to issue such permits subject to conditions and limitations; and mandates sharing of information relating to permitted vessels with the FAO. The HSFCA also prohibits use of high seas fishing vessels in contravention of international conservation and management measures recognized by the United States, or in a manner that would violate a permit condition.

Shark Conservation Act of 2010. Enacted January 4, 2011, the SCA, 16 U.S.C. 1801 note, amended the Moratorium Protection Act to promote adoption by RFMOs of shark conservation measures, including banning removal of any of the fins of a shark and discarding the carcass at sea. The Act amended the definition of IUU fishing with an explicit reference to violation of international shark conservation measures, and provides for identification of a nation for activities related to shark conservation.

Lacey Act. The Lacey Act, 16 U.S.C. 3371-3378, prohibits the import, export, transport, sale, possession, or purchase in interstate or foreign commerce of any fish or wildlife taken, possessed, transported, or sold in violation of any U.S. state law or regulation or of any foreign law. The two-part prohibition requires evidence of a violation of domestic or foreign law, and also evidence of trafficking. NMFS has used the law to prosecute foreign individuals who import fish caught without authorization in another country's EEZ.

Marine Mammal Protection Act. A goal of the MMPA, 16 U.S.C. 1361 *et seq.*, is to reduce the incidental kill or serious injury of marine mammals in the course of commercial fishing to insignificant levels, approaching zero. The Act prohibits "taking" (actual or attempted harassment, hunting, capture, or killing) and importation into the United States of marine mammals except where explicitly authorized. The MMPA also bans the importation of fish caught with commercial fishing technology that results in the incidental kill or serious injury of marine mammals in excess of U.S. standards.

Endangered Species Act. This Act, 16 U.S.C. 1531 *et seq.*, provides for the conservation of species that are in danger of extinction throughout all or a significant portion of their range. The Act lists species as either "threatened" or "endangered." When a species is endangered, it is protected from being "taken" through harassment, harm, injury, pursuit, hunting, killing, capturing, or collection. The Act also provides for U.S. implementation of limitations on trade of species listed under CITES.

International Dolphin Conservation Program Act. This Act, Public Law 105-42, amended the MMPA to provide that nations whose vessels fish for yellowfin tuna with purse seine nets in the Eastern Tropical Pacific are permitted to export such tuna to the United States only if the nation provides documentary evidence that it participates in the International Dolphin Conservation Program and is a member (or applicant member) of the IATTC, is meeting its obligations under the Program and the IATTC, and does not exceed certain dolphin mortality limits.

Shrimp-Turtle Law (Section 609 of P.L. 101-162). This law, 16 U.S.C. 1537, requires the United States to embargo wild-caught shrimp harvested with commercial fishing technology, such as trawl nets, that may adversely affect sea turtles. The import ban does not apply to nations that have adopted sea turtle protection programs comparable to those of the United States.

Nations seeking to import shrimp must be certified by DOS as meeting the law's requirements on an annual basis.

Pelly Amendment. The Pelly Amendment to the Fishermen's Protective Act of 1967, 22 U.S.C. 1978, provides for the possibility of trade-restrictive measures, as described in Part I.D.

SUMMARIES OF RECENT ENFORCEMENT CASES WITH AN INTERNATIONAL NEXUS

This section summarizes recent U.S. enforcement cases involving an international nexus such as IUU fishing by a foreign-flagged vessel, international trafficking in seafood illegally harvested or labeled, and U.S. assistance with another nation's investigation of a fisheries violation. NOAA, the USCG, and the U.S. Department of Justice are actively engaged around the country and overseas in monitoring fishing activity for a number of ecologically and economically valuable marine species. These efforts in combating IUU fishing and PLMR bycatch not only help to protect global fish stocks and other marine resources, but also preserve the integrity of the U.S. domestic fish market and the safety of the U.S. food supply. Some of the more significant cases since January 2013 are outlined below:

- A multiyear investigation involving NOAA, the U.S. Fish and Wildlife Service, Environment Canada, and the U.S. Department of Justice has dismantled an international ivory smuggling ring. Three U.S. citizens and a Canadian conspired for nearly a decade to transport narwhal tusks from Canada to the United States by smuggling them across the border, in violation of the MMPA. One defendant purchased tusks that he knew had been illegally imported; he was convicted February 14, 2014, of Lacey Act, money laundering, and smuggling offenses by a federal jury in Bangor, Maine. Earlier in 2014, two other defendants pleaded guilty to conspiracy and other offenses, one in Bangor and one in Anchorage. The three Americans had sold ivory through internet and direct transactions. A Canadian citizen supplied and transported the narwhal tusks into the United States by driving across the Canada/Maine border with a hidden compartment under his vehicle. A former Mountie, he is currently going through the extradition process in Canada.

- The USCG Cutter *Morganthau*, aided by Canadian air patrol, Japanese fisheries observers, and Chinese shipriders, investigated two apparently Stateless transshipment vessels and apprehended a Chinese driftnet vessel in the North Pacific in May 2014. This series of significant enforcement actions is described in detail in Part VI.C.1.
- In August 2013, NOAA successfully prosecuted two separate civil penalty cases against U.S. owners, operators, and fishing masters of purse seine vessels used in violation of regulations implementing CMMs of the WCPFC. In the first case, the *American Triumph* conducted six sets on or within one nautical mile of a fish aggregating device, which had been deployed during a closure prohibiting use of such devices. The violations resulted in a civil penalty of $562,068. In the second case, five separate fishing vessels were shown to have made 10 sets on or within one nautical mile of a fish aggregating device, and to have deployed two such devices during the closure. Those vessels also set their purse seine nets on whales five times, which is a violation of the MMPA. The administrative law judge assessed a civil penalty of $953,054. In both cases, the violations were reported by observers from the Forum Fisheries Agency as part of the compliance scheme of the WCPFC.
- Three convicted smugglers of South African rock lobster were ordered on June 14, 2013, to pay restitution under the Lacey Act of about $22.5 million to the Government of South Africa. A U.S. District Judge had ruled in 2007 that restitution was not available under the Lacey Act, but the Second Circuit Court of Appeals overturned that ruling. Arnold Bengis, Jeffrey Noll, and David Bengis had engaged in an elaborate scheme to catch illegal quantities of rock lobster and other fish and to import them to be processed in its American factory. The three men served prison sentences and suffered forfeitures in the United States; a South African prosecution resulted in further fines and forfeitures.
- On August 1, 2014, True Nature Seafood, LLC, a Miami-based company, was convicted of false labeling of seafood products under the Lacey Act. The company had imported the fish from Chile, changed the label from steelhead trout to salmon, and shipped the product to customers in the United States and abroad. The U.S. District Judge imposed fines totaling $1 million, placed the company on probation for five years, required it to implement an Environmental Compliance Plan, and order the forfeiture of more than $400,000, the

proceeds from the sale of toothfish previously seized by NOAA for a separate offense.

- On June 12, 2013, the United States District Court for the Northern District of New York authorized civil penalties totaling $35,000, forfeiture of $96,014 in proceeds from the sale of 9,600 pounds of Patagonian toothfish, and injunctive relief against Pescanova, Inc. These sanctions were based upon the company's exportation and subsequent re-exportation of 9,600 pounds of Patagonian toothfish (also known as Chilean sea bass) in 2010. That trade violated U.S. regulations implementing CCAMLR's CDS.
- The 2013 Report included a case with very large fines and forfeitures resulting from a scheme to import protected black coral into the United States. On February 7, 2013, the former president and CEO of GEM Manufacturing LLC was sentenced for his role in the scheme. The U.S. District Judge imposed a fine of $918,950 and sentenced Ashu Bhandari to a month in jail, followed by a month of home confinement, a year of supervised release, and 300 hours of community service. The defendant is also banned from any business venture involving coral or coral products, and must pay $229,587 to the University of the Virgin Islands for research and protection of black corals.
- On April 18, 2013. the USCG and a shiprider from the Federated States of Micronesia onboard the *Reuben James* boarded the *Hao Xiang* in the Micronesian EEZ. The boarding identified two violations: no documentation of a Micronesia-required VMS unit and no license to transship in the EEZ. The FSM shiprider issued fines of $1,500.
- A total of 15 suspected incursions were detected by electronic sources by the South Korean vessel *Joon Sung Ho* along the Maritime Boundary Line in the Bering Sea in July 2013. The vessel had been permitted to fish inside the Russian EEZ. The USCG in Juneau worked with Russian counterparts at the Kamchatka Border Guard to report incursions and request assistance. After warnings by both Russian and U.S. agencies, it became apparent that the vessel had incorrect coordinates for the boundary. The USCG provided correct coordinates, after which the vessel made no further incursions.

ANNEX 3: SEABIRD BYCATCH ISSUES

Seabirds fall within the definition of international living marine resources under the Moratorium Protection Act, but not within the definition of protected marine living resources. Section 316 of the MSA highlights the need for the Secretary of Commerce to work cooperatively with the Secretary of the Interior, with regional fishery management councils, and within international organizations to seek ways to mitigate seabird bycatch. NMFS has pushed hard internationally for action to protect seabirds, particularly measures to mitigate seabird bycatch in fisheries.

The Agreement on the Conservation of Albatrosses and Petrels (ACAP) coordinates international activity to mitigate known threats to albatross and petrel populations. The ACAP treaty was submitted to the Senate in 2008 for its advice and consent to ratification; draft implementing legislation was submitted to Congress in 2009. The United States participates in ACAP meetings as an observer due to its interest in seabird conservation and its status as a range State under ACAP. The United States contributed information and suggestions during Advisory Committee meetings of ACAP held in May 2013 and September 2014. The United States contributed to revision of best practice advice to incorporate side-setting measures that have been used effectively by some vessels in the Hawaii longline fishery, brought forward experience with electronic monitoring, and made proposals toward developing seabird bycatch mitigation measures.

In 2014, the United States published a report on implementation of its National Plan of Action for Reducing the Incidental Catch of Seabirds in Longline Fisheries. The report highlights advances made by the United States toward the objectives of the 2001 Plan. Since 2001, the United States has improved research, outreach, education, and domestic management of seabird bycatch, resulting in a significant decrease in such bycatch in its domestic fisheries. The report is available at http://www.nmfs.noaa.gov/ia /species /seabirds/seabirds.html.

Several RFMOs have considered or taken action concerning seabirds in 2013 and 2014:

CCAMLR. Observed seabird bycatch in the Convention Area is near zero in the legal fishery outside of the French EEZ. Seabird bycatch within the French EEZ, historically a problem, continues to decline significantly each year due to improved mitigation and management measures. Incidental catch was reported at 141 seabirds for the 2012–2013 fishing season, the lowest level recorded yet. In 2013, CCAMLR began implementing an evaluation procedure to examine compliance by member vessels with requirements, including those related to mitigation of seabird bycatch. During the 2013 meeting, one United Kingdom vessel was found to have discharged offal contrary to a prohibition. The scuppers of the vessel were subsequently modified to prevent such discharge. During the 2014 meeting, a Chilean vessel was found to have set longline gear after morning twilight, contrary to a requirement for setting gear only at night, which resulted in bycatch of 74 white-chinned petrels. The vessel had not deployed streamer lines during one set. Chile is investigating these incidents of non-compliance.

ICCAT. In 2011, ICCAT agreed to a supplementary recommendation on reducing incidental bycatch of seabirds in longline fisheries. This measure strengthens the 2002 requirements for mitigation and moves toward implementation of best practices. The 2011 recommendation requires use of two of three measures (night setting, branch line weighting, and bird scaring lines) for vessels fishing south of 25° S, and recommends voluntary use of the measures in the Mediterranean and other areas as appropriate. The recommendation also establishes binding minimum technical standards for each of the measures and provides additional technical guidance for design and deployment of tori lines. The recommendation went into effect in July 2013. The SCRS will undertake another fishery assessment in 2015 to evaluate the efficacy of the measures.

IATTC. In 2011, the IATTC adopted a measure to mitigate the effect of fishing on seabirds. In 2014, the IATTC scientific staff recommended revising the measure to be consistent with current advice regarding seabird mitigation techniques. The staff also recommended taking note of updated seabird density information and considering expansion of the area of application to include additional waters in the North Pacific (Mexico's EEZ, currently exempt). No action was taken on the recommendations.

IOTC. In 2012, the IOTC adopted a resolution to reduce the incidental bycatch of seabirds in longline fisheries, which entered into force on July 1, 2014. This measure is substantially the same as the ICCAT recommendation described above. The Scientific Committee will evaluate the resolution prior to the 2016 meeting.

SPRFMO. In January 2014, SPRFMO adopted a CMM requiring action to minimize bycatch of seabirds in demersal longline and trawl fisheries by July 31, 2015. For demersal longline vessels, the measure requires a combination of bird scaring lines, line weighting, and night setting. If there is observer coverage of at least 10 percent and a recorded seabird mortality rate less than 0.01 birds/1,000 hooks, a member may choose to require its vessels to apply only one of the three measures noted previously and ensure a minimum of 10 percent observer coverage that is adequately representative of the spatial and temporal distribution of the fishing fleet. Any member that applies the second option but exceeds the rate of 0.01 bird/1,000 hooks is subject to additional requirements. Trawl vessels are required to deploy devices to deter birds away from warp cables and net monitoring cable, either two bird-scaring lines or, where operational practices prevent their effective deployment, a bird baffler. Vessels are also to manage discharge of offal and other biological materials to avoid attracting seabirds to the vessel.

End Notes

[1] The MSRA's use of the term is more circumscribed and complicated; see Part II.A for definitional details. Section 402 of the MSRA contains a finding that international cooperation is necessary to address "illegal, unreported, **and** unregulated fishing" (emphasis added). On the other hand, Section 403 of the MSRA, which establishes the standards for identification and certification of nations whose vessels engage in IUU fishing, uses a disjunctive formulation of the term, referring to nations whose vessels are engaged in "illegal, unreported, **or** unregulated fishing" (emphasis added). The FAO and other international bodies generally employ the conjunctive formulation of the term in publications, plans of action, and related materials. In this report, we use the phrase "IUU fishing" without indicating whether the conjunctive or disjunctive formulation is intended, but with the understanding that where identification and certification determinations are at issue under the MSRA, the term is to be understood and employed in the disjunctive. We do not intend any particular legal meaning or consequence to flow from the use of the term in this report.

[2] The FAO reported that IUU fishing has escalated over the past 20 years, citing rough estimates that the annual quantity of fish harvested illicitly is 11 to 26 million tons, worth between $10 and $23 billion. FAO, "The State of World Fisheries and Aquaculture," Rome, 2014, p. 84.

[3] Bycatch of seabirds could not serve as the basis for identification of a nation under the PLMR provisions of the MSRA, but violations of seabird measures adopted by RFMOs of which the United States is a member could serve as the basis for identification under the Act's IUU fishing provisions.

[4] The prohibition does not apply to individuals engaged in commercial fishing for smooth dogfish, under certain conditions and circumstances. On August 7, 2014, NMFS published Draft Amendment 9 to the 2006 Consolidated Highly Migratory Species Fishery Management Plan, which considers smoothhound shark management measures including the SCA exception for smooth dogfish (79 Federal Register 46217). A final rule for this action is expected in 2015.

[5] The 2011 Report (at page 44) describes a NMFS advance notice of proposed rulemaking to implement that provision with regard to the import of fish and fish products. NMFS expects to publish the proposed rule in 2015.

[6] During the 113th Congress, the Senate Committee on Commerce, Science, and Transportation reported out S. 269, which would have made the following changes in Section 609(a):

- A nation would be identified if it is engaged, or has been engaged at any time during the preceding three years, in IUU fishing that undermines the effectiveness of measures required under the relevant RFMO.
- A nation would be identified if it is violating, or has violated at any time during the preceding three years, conservation and management measures required under an international fishery management agreement to which the United States is a party and the violations undermine the effectiveness of such measures.
- A nation would be identified if it is failing, or has failed at any time in the preceding three years, to effectively address or regulate IUU fishing.

The House Committee on Natural Resources reported favorably on H.R. 69, which contained similar amendments to Section 609(a), on September 18, 2014.

[7] The NPFC measures, at this point, are non-binding.

[8] The gap between the convention areas of the North Pacific and South Pacific RFMOs.

[9] Large portions of the Indian Ocean are covered by the Southwest Indian Ocean Fisheries Agreement, which came into force in 2012. The Commission established under that agreement, however, has not yet established conservation measures to control bottom fishing.

[10] S. 269 and H.R. 69 also would have expanded the time period for identifying nations for bycatch activities in Section 610(a)(1) to three years.

[11] Sharks that are shared PLMRs and are incidentally caught in another nation's EEZ would be considered as a basis for identification under Section 610(a)(1).

[12] S. 269 and H.R. 69 proposed no change to Section 610(a)(2) affecting the time frame.

[13] The sources of information on Colombia fishing activities are "Possible non-compliance cases of IATTC resolutions, 2013 REVISED" and communications from NOAA's National Data Buoy Center.

[14] The sources of information on Ecuador fishing activities are "Possible non-compliance cases of IATTC resolutions, 2013 REVISED" and communications from NOAA's National Data Buoy Center.

[15] The sources of information on Mexico's fishing activities are 24 case package reports from USCG District Eight transmitted to Director General de Inspeccion y Vigilancia, Comision Nacional de Acuacultura y Pesca, dated between February 21, 2013 and May 29, 2014.

[16] 78 Federal Register 3338, January 16, 2013.

[17] The source of information on Nicaraguan fishing activities is "Possible non-compliance cases of IATTC resolutions, 2013 REVISED."

[18] The sources of information on Nigerian fishing activities are various Communication Circulars from Australia, France, New Zealand, and Japan in 2013 and 2014.

[19] http://www.seashepherdglobal.org/news/latest-news/sea-shepherd-intercepts-toothfish-poachers-in-thesouthern-ocean.html (accessed December 18, 2014).

[20] The sources of information on Portuguese fishing activities are NAFO STACTIC Working Paper 14/16, Table C-6: Details of Apparent Infringements detected in 2013, GSF/14-173, 23 June 2014; Fisheries and Oceans Canada NAFO Citations web page: http://www.dfo-mpo.gc.ca/international/mcscitations-eng.htm, accessed August 14, 2014.

[21] The sources of information on Belize fishing activities are "Possible non-compliance cases of IATTC resolutions, 2013 REVISED" and communications from NOAA's National Data Buoy Center.

[22] The sources of information on Costa Rican fishing activities are from USCG boarding reports.

[23] The sources of information on Ghanaian fishing activities are "ICCAT Report for biennial period 2012 –2013 Part II (2013) – Vol. 1, 2014" and "African Winds Report, Ghana and Benin, IMCS Network, October 2013."

[24] The sources of information on Guatemalan fishing activities are from USCG boarding reports.

[25] The sources of information on Spanish fishing activities are NAFO STACTIC Working Paper 14/16, Table C-6: Details of Apparent Infringements detected in 2013, GSF/14-173, 23 June 2014; Fisheries and Oceans Canada NAFO Citations web page: http://www.dfo-mpo.gc.ca/international/mcs-citationseng.htm, accessed August 14, 2014.

[26] The term "international living marine resources," as described in this sentence, is much more inclusive than the term "protected living marine resources." The latter includes only non-target species protected under U.S. law or international agreement that, except for sharks, are not managed under the MSA, the Atlantic Tunas Convention Act, or any international fishery management agreement.

[27] South Korean authorities inspected the two transshipment vessels in the port of Busan. Information on the transshipment vessels was also passed to Russian officials.

In: International Fisheries Management
Editor: Marlin D. Anderson
ISBN: 978-1-63483-170-3

Chapter 2

REAUTHORIZATION ISSUES FOR THE MAGNUSON STEVENS FISHERY CONSERVATION AND MANAGEMENT ACT*

Harold F. Upton

SUMMARY

The Magnuson-Stevens Fishery Conservation and Management Act (MSFCMA governs the management and conservation of commercial and recreational fisheries in U.S. federal waters (3-200 nautical miles from shore). The MSFCMA was last reauthorized and extensively amended in 2006 (P.L. 109- 479). Although the authorization of appropriations under the MSFCMA expired at the end of FY2013, the act's requirements continue in effect and Congress has continued to appropriate funds to administer the act. Historically, reauthorization has also provided the opportunity to introduce significant amendments to the act.

During the first decade after the act was passed in 1976, fishery policy focused on controlling and replacing foreign fishing and developing U.S. fisheries in the newly declared 200-mile Fishery Conservation Zone. After that time, new issues emerged, including recognition of the need to sustain fish populations and respond to overfishing while attempting to satisfy the economic and social needs of recreational and commercial fishermen and

* This is an edited, reformatted and augmented version of a Congressional Research Service publication R43565, prepared for Members and Committees of Congress, dated May 22, 2014.

fishing communities. Achieving this balance is closely related to allocating federal fishery resources among different users, developing and supporting existing management institutions, and investing in management and research.

This report covers issues that have been identified during congressional hearings and in legislation introduced during the last three Congresses. Although most issues are not new, they have evolved with changes to the statute, regulations, and fishery management plans. Major issues include (1) flexibility in rebuilding overfished fisheries, (2) annual catch limits, (3) uncertainty and data needs; (4) catch shares (limited access privilege programs), (5) management process and decision making, (6) bycatch, and (7) environmental quality. A variety of other issues are also covered in this report.

Most of these issues are part of a system of linked elements including ecosystems (fish populations and biophysical elements of the environment), fishing (commercial and recreational fishermen, processors, and other related fishing businesses), management (managers, scientists, and the regulatory system), fishing communities (other related businesses and coastal residents), and markets (wholesale, retail, restaurants, and consumers). Often a change in one element affects other elements. For example, requirements to stop overfishing that use restrictive catch limits may rebuild fish populations, but may also result in short-run harm to fishing businesses and coastal communities.

During the last three Congresses, over 30 different bills have been introduced to amend portions of the MSFCMA. These bills have covered a wide variety of topics, ranging from proposals to change management for specific fisheries or regions, to general changes to the management process such as requirements of fishery management plans. Oversight hearings concerning MSFCMA reauthorization have been held by the House Committee on Natural Resources and by the Senate Committee on Commerce, Science, and Transportation. On December 18, 2013, the chairman of the House Committee on Natural Resources released a draft that included many elements of previously introduced bills. In early April 2014, the Senate Committee on Commerce, Science, and Transportation also released a reauthorization draft.

INTRODUCTION

The Magnuson Stevens Fishery Conservation and Management Act (MSFCMA, 16 U.S.C. §§1801 et seq.) governs the management and conservation of commercial and recreational fisheries in U.S. federal waters (3-200 nautical miles from shore). Although the MSFCMA has been amended a least 30 times since it was enacted in 1976, the act has retained many of its original elements. The act decentralized the federal management process by setting up regional fishery management councils and requiring extensive public comment during the development of fishery management plans. The MSFCMA also has evolved with changes to fishery resources, the U.S. fishing industry, recreational fishing, and seafood markets. Generally, the challenges of fisheries management have shifted from developing fisheries to addressing conservation of fish populations and the marine environment.[1]

The MSFCMA was last reauthorized and extensively amended in 2006 (P.L. 109-479). Although the authorization of appropriations under the MSFCMA expired at the end of FY2013, the act's requirements continue in effect and Congress has continued to appropriate funds to administer the act. Historically, reauthorization has also provided the opportunity to introduce significant amendments to the act. As Congress considers reauthorization, it faces the ongoing challenge of balancing utilization and conservation of fish populations. Some of the main questions revolve around stopping overfishing and rebuilding fish stocks while maintaining the well-being of fishermen and fishing communities. Related management issues involve the quality of data and stock assessments used for managing fisheries and the amount of flexibility allowed in the management process.

During the last three Congresses, a number of bills have been introduced to address these issues and other concerns related to fisheries management. Oversight and reauthorization hearings have been held by the Senate Committee on Commerce, Science, and Transportation and the House Committee on Natural Resources. In December 2013, the Chairman of the House Committee on Natural Resources released a reauthorization discussion draft.[2] The draft includes several sections that reflect topics covered by bills introduced in previous sessions of Congress. In early April 2014, the Senate Committee on Commerce, Science, and Transportation also released a draft to stakeholder groups.[3]

ISSUES FOR CONGRESS

An ongoing issue for managers, fishermen, and environmentalists is the balance between conservation and utilization of fish populations. Although there is general agreement that stocks[4] should not be overfished and overfished stocks should be rebuilt, questions remain with regard to the timing of management actions, the choice of management objectives, how stock management objectives should be achieved, and the information needed to make these determinations. Several interrelated issues have emerged during the ongoing debate over requirements to use annual catch limits (ACLs) and to rebuild fish populations. General categories of issues include (1) providing for greater flexibility during stock rebuilding; (2) incorporating new data and uncertainty when using ACLs; (3) improving the decision-making process; (4) establishing limited access privileges; and (5) reducing bycatch. Meanwhile, managers must also contend with environmental factors over which they often have no control such as climate change and the loss and degradation of fish habitat. Decreasing environmental quality may be the greatest long-term threat to the productivity of many fish populations.

Flexibility in Ending Overfishing and Rebuilding Overfished Fisheries

Flexibility in rebuilding overfished fisheries has become one of the main issues of the current MSFCMA reauthorization debate. The MSFCMA was amended to require development and use of ACLs to end overfishing for all federally managed stocks in the 2010 fishing year.[5] Overfished stocks are required to have rebuilding plans that adhere to a 10-year timeframe (with some exceptions). Previously, managers often delayed action or set indirect controls on fishing such as gear restrictions or closed areas, sometimes with disastrous results for the fishery. RFMCs and National Marine Fisheries Service (NMFS) are now required to set ACLs (quotas) within specific biologically determined levels.

Fishery management plans (FMPs) must be consistent with the 10 national standards in Section 301(a) of the MSFCMA. Council members must address the national standards as they develop FMPs and, when considering approval, the Secretary of Commerce determines whether FMPs are consistent with these national standards. The national standards cover a broad range of basic fishery management objectives. The first National Standard Section 301(a)(1) states: "Conservation and management measures shall prevent overfishing while achieving, on a continuing basis, the optimum yield from each fishery for the United States fishing industry."

Provisions of the MSFCMA enacted during the 1996 reauthorization and amended during the 2006 reauthorization added specific requirements to end overfishing and to rebuild overfished fish stocks. To implement these requirements, the MSFCMA directed the National Marine Fisheries Service of the National Oceanic and Atmospheric Administration to update National Standard 1 guidelines by 2008 to provide guidance for establishing annual catch limits (ACLs) and related biological benchmarks.[6]

On January 16, 2009, NMFS issued guidelines that describe fishery management approaches to meet the objectives of National Standard 1 with emphasis on new requirements to end overfishing and rebuild overfished stocks. While implementing ACL requirements, NMFS has identified a number of issues that may require additional revisions to the National Standard 1 guidelines. On May 3, 2012 NOAA published an advance notice of proposed rulemaking to request public comments on potential adjustments to the guidelines.[7] No further action has been taken since the Summary of Comments on Advanced Rulemaking was published.

The terms overfishing and overfished are often confused and assumed to occur together, but this is not necessarily the case. According to the National Standard Guidelines:

> overfishing occurs whenever a stock or stock complex is subjected to *a rate or level of fishing mortality* that jeopardizes the capacity of a stock or complex to produce Maximum Sustainable Yield (MSY) on a continuing basis. A stock or stock complex is considered overfished when its *biomass has declined below a level* that jeopardizes the capacity of the stock or stock complex to produce MSY on a continuing basis.[8]

Overfishing occurs when the rate of removals (catch or harvest) is high relative to the size of the fish stock; fish stocks are overfished when their biomass is relatively low. At certain points during rebuilding, removals may be low (no overfishing), but the stock is still overfished (its biomass is not yet rebuilt). Conversely, removals may be high and overfishing may be occurring, but the stock biomass has not declined to the point at which the stock is considered to be overfished.

Issues

Stock rebuilding has become controversial because rebuilding timeframes have required strict constraints on harvest levels. Some have questioned whether greater flexibility in determining the length of stock rebuilding periods could increase economic benefits from the fishery. Rebuilding plans with greater flexibility could also contribute to other fishery management goals such as needs of fishing communities (National Standard 8).

Overfishing has been arrested in most U.S. fisheries and progress has been made in rebuilding many others. As of December 31, 2013, of the 300 stocks with a known overfishing status, 28 stocks were subject to overfishing and of the 230 stocks with a known overfished status, 40 stocks are classified as overfished.[9] NOAA also reported that 34 fish stocks have been rebuilt since 2000.

According to a recent National Research Council (NRC) study, "fishing mortality of stocks placed under rebuilding plans has generally been reduced and stock biomass has generally increased following reductions in fishing mortality."[10] However, these improvements have sometimes come at a cost to commercial and recreational fishermen and associated fishing communities, and in some cases stocks have not responded to management actions as managers anticipated. The NRC study attributed some of the mixed performance of rebuilding plans to scientific uncertainty and a mismatch between policy makers' expectations for scientific precision and the inherent limits of science. The NRC study adds that mixed outcomes of rebuilding plans have "added to concerns with the significant social and economic costs associated with implementation of time-constrained rebuilding plans."

Stakeholder Views

Fishermen and fishing communities sometimes suffer from economic and social effects of harvest restrictions needed to satisfy MSFCMA overfishing and stock rebuilding requirements. Many question whether these requirements adequately address the complexities and uncertainties associated with managing fish stocks. Often fishermen express doubt over the efficacy of fish population assessments used for developing management measures because of data constraints and inadequate population models. Furthermore, they refer to studies showing that other factors, often outside the immediate control of fisheries managers, such as environmental conditions and the quality of fish habitat, also affect fish population abundance.

Others, including environmentalists and fishery managers, counter that overfishing and previous management failures illustrate the need to maintain established fish stock rebuilding schedules. They emphasize that relatively short-term sacrifices today will result in long-term economic gains to recreational and commercial fishermen in the future. They point to stocks that have been rebuilt since 2000 and cite notable examples of fully rebuilt stocks such as Northeast scallop, Mid-Atlantic bluefish, and Pacific lingcod.

Environmentalists also have asserted that many species could be rebuilt within 5 years and that the 10-year requirement is a balance between biology of most species and short-term concerns of some managers and fishermen.[11] Exceptions have been provided for many species that have rebuilding timeframes greater than 10 years because of their life history.[12] Others contend that the 10-year rebuilding timeframe is arbitrary and that the lack of flexibility prevents regulators from pursuing a more balanced approach.[13]

Economic Concerns

Economists and social scientists have questioned whether rebuilding plans that ignore the unique characteristics of each fishery, such as social and economic considerations, may result in significant loss of social welfare.[14] They contend that stock rebuilding targets should not be based solely on biological factors. Depending on the productivity of the stock, characteristics of the fishery, and the discount rate, extending the rebuilding timeframe may increase economic benefits. Moreover, economic and social analysis can be useful when developing management measures used to achieve management objectives. Recognition of economic factors can ensure that the least costly or least socially disruptive management alternatives have been considered. Some social scientists argue that economic and social analyses are often incorporated after biological objectives have been established. For these reasons some social scientists have stressed the need to integrate social and economic elements from the beginning of the process.

Increasing management flexibility also might improve short-term economic returns and lessen immediate social impacts on commercial and recreational fishermen. When factors outside of the control of fisheries managers occur such as environmental changes, management flexibility also might lessen the severity of economic and social disruption to fishermen until conditions improve. It has been reported that most fish stocks experience productivity regime shifts related to natural environmental fluctuations.[15] However, for some fisheries existing flexibility may be adequate and in these cases greater flexibility could delay or stop progress toward stock rebuilding and increase long-term social costs. Unfortunately, developing specific rules for all fisheries is difficult if not impossible because fisheries are diverse with regard to the biology of target species, technology of harvesting strategies, and socioeconomic elements of related communities.

Others have expressed concerns that managers have lost sight of the original fisheries management goals related to employment, food supply, revenue, and recreational opportunities. According to recent congressional testimony by Dr. Ray Hilborn, a consequence of reducing overfishing is to underutilize other fish stocks.[16] The testimony asserts that 77% of stocks are underfished[17] and 30-48% of U.S. potential yield is lost by underfishing. At the same time 22% of stocks that are overfished only constitute 1-3% of potential yield because they contribute to relatively small fisheries. He postulates that that lack of fishing effort associated with underfishing occurs for a number of reasons such as the lack of markets, but one of the primary factors is because precautionary regulations have been imposed to prevent overfishing. However, economic profitability of commercial harvesting is maximized at fishing effort levels below those which produce MSY.[18] Consideration of economic efficiency under national standard 5 is another management objective that could be met by decreasing fishing effort to maximize net benefits instead of maximizing production. Unfortunately, satisfying management objectives is often more complex because of other elements of the system that require consideration such as recreational fishing allocations and benefits, processing and marketing sectors, and the well-being of fishing communities.

The requirements to stop overfishing and rebuild stocks by using ACLs may improve fishing in the long run, but they also affect the allocation of fishing opportunities, catch, and benefits among fishermen and related businesses. Questions arise with respect to when benefits will accrue to fishermen and who will ultimately benefit when stocks have been rebuilt. Distributional issues may exist among different commercial gear types, commercial and recreational fishermen, and ports or communities. The effects of stock rebuilding may vary across different segments of the fishing industry such as support services, harvesters, processors, wholesalers, and retailers.[19] There is also a temporal dimension to allocation for many fisheries because of the potential decoupling of present costs and future benefits. Often there are few if any guarantees that those who endure the immediate costs of ACLs and rebuilding programs will benefit in the future because of the weak nature of property rights in many fisheries[20] and the inability of some fishermen to remain in the fishing industry when economic returns decline.

Specific segments of fishing fleets, especially small-scale or traditional fishermen,[21] and fishing communities may be affected disproportionately by requirements to end overfishing and by stock rebuilding programs. Like many segments of the U.S. economy, the fishing industry is changing, in this case due to a mixture of factors related to technology, social views, and resource limits.[22] The current emphasis of fisheries management on long-term sustainability has resulted in less flexibility for fishermen; fishermen often find regulations restrict their access to fisheries, especially in cases where strict stock rebuilding is required. In some cases smaller vessels and specific coastal communities have been affected disproportionately because of their scale. For example, requirements to carry observers are disproportionately costly for smaller businesses and alternatives such as more distant fishing grounds may not be accessible by smaller vessels.

Multispecies Fisheries

Multispecies fisheries are often difficult to manage because the fishery may consist of both healthy and overfished stocks. Harvesting one stock at its optimum yield may result in overfishing of another stock when the two stocks are caught together as part of one fishery or when one of the stocks is caught as bycatch in another fishery. Fishing on healthy stocks is sometimes constrained by restrictions to promote rebuilding of another stock(s) that have been identified as overfished.[23] When the quota of the overfished stock is reached, the entire fishery may be closed or curtailed. For example, over 90 species of groundfish are managed by the Pacific Regional Fishery Management Council, but fishing regulations are driven by eight species that are under rebuilding plans. This has led to under-fishing of healthy stocks and lost yield from underfishing amounts to 30% of total sustainable yield.[24] This infers that greater flexibility in rebuilding timelines would allow for greater profits in this fishery.

Most observers would agree that a biological biomass threshold is necessary to avoid depletion of overfished populations or, in the worst case, to avoid extinction. However, some question whether there should be greater flexibility in setting the level of stock biomass thresholds for weaker stocks in multispecies fisheries. The NMFS Guidelines attempt to address this issue with the mixed stock exception.[25] The purpose of the mixed stock exception is to provide managers with a means to achieve Optimum Yield (OY) for some species while allowing overfishing of other species. According to the 2009 guidelines, the mixed stock exception may allow overfishing, but not if the stock is overfished or if the stock would be decreased to levels which would require stock rebuilding.[26] Generally, it appears that as currently specified in the guidelines the mixed stock exception could only be used in limited circumstances and not in cases where stock have already been overfished.

Bills Introduced During the 112th Congress

Several bills introduced during the 112th Congress (H.R. 1646, H.R. 3061, H.R. 6350, and S. 632) would have added similar provisions to increase management flexibility. These bills would have amended Section 304(e)(4)(A)(i) of MSFCMA by changing the requirement from rebuild as soon as "possible" to a requirement to rebuild as soon as "practicable." The following exceptions also would have been added to the current 10-year rebuilding requirement.

(II) the Secretary determines that such 10-year period should be extended because the cause of the fishery decline is outside the jurisdiction of the Council or the rebuilding program cannot be effective only by limiting fishing activities;

(III) the Secretary determines that such 10-year period should be extended to provide for the sustained participation of fishing communities or to minimize the economic impacts on such communities, provided that there is evidence that the stock is on a positive rebuilding trajectory;

(IV) the Secretary determines that such 10-year period should be extended for one or more stocks of a multi-species fishery, provided that there is evidence that those stocks are on a positive rebuilding trajectory;

(V) the Secretary determines that such 10-year period should be extended because of a substantial change to the biomass rebuilding target for the stock of fish concerned after the rebuilding plan has taken effect; or

(VI) the Secretary determines that such 10-year period should be extended because the biomass rebuilding target exceeds the highest abundance of the stock of fish during the 25-year period preceding the date the rebuilding plan has taken effect and there is evidence that the stock is on a positive rebuilding trend.

The Secretary also would have been required to review factors other than commercial and recreational fishing that may contribute to the overfished status of a given stock of fish. Examples include environmental harm caused by commercial, residential, and industrial development, and agriculture in coastal areas, predator-prey relationships of target and related species, and other environmental and ecological changes to marine conditions. The rebuilding time period would be limited to the sum of the initial 10-year period, the time required to rebuild the stock without any fishing mortality, and the mean generation time of the stock.

Uncertainty, Data, and Annual Catch Limits

A provision added to the MSFCMA in 2006 requires fishery management plans to include a mechanism for specifying annual catch limits (ACLs) at a level where overfishing does not occur. The ACL requirements took effect in 2010 for fisheries subject to overfishing and in 2011 for all other fisheries. If ACLs are set at appropriate levels as required by MSFCMA, this action would end overfishing for all federally managed fish stocks. ACLs are defined in NMFS guidelines as the level of annual catch of a stock or stock complex that may not exceed allowable biological catch and serves as the basis for using accountability measures (AMs). AMs are actions taken to ensure that rebuilding will continue when adjustments are needed relative to the ACL. AMs include measures taken during the season to prevent the ACL from being exceeded or adjustments in the next fishing year to compensate for overages if the ACL was exceeded.

Uncertainty

The complexity of marine ecosystems and fisheries not only make it difficult to determine ACLs and target stock levels, but because of the system's dynamic nature, benchmarks and forecasts are constantly changing. The NOAA Guidelines identify two types of uncertainty – management uncertainty and scientific uncertainty.[27] Management uncertainty occurs because of the lack of information on actual catch due to illegal activity, late reporting of catch, misreporting catch, or non-reporting of bycatch. Landings data are rarely complete, especially for those fisheries with significant discards or a large recreational component. In these cases, managers have insufficient information to know whether an ACL has been reached and to make related management decisions such as slowing fishing effort or closing the fishery.

Scientific uncertainty is the uncertainty associated with the estimates of stock biomass and fishing mortality rates. Scientific uncertainty may occur for different reasons including limited biological data for many fisheries and inadequate stock assessment models. Furthermore, even for the most closely studied stocks, spawning success and future recruitment to the population are difficult to predict. The relationship between the abundance of spawning adults and recruitment (off-spring entering the populations) is confounded by biological and environmental factors. Assessments are also out of date by the time they are completed. First there is a lag between the time data are collected and the time taken to compile data and complete the assessment. Assessments are usually undertaken every three to five years because of funding constraints.

Sometimes unpredictable and significant changes may occur before updates can be undertaken. In addition, factors affecting management uncertainty such as mischaracterization of catch may also increase scientific uncertainty. Some level of uncertainty is inevitable because of the nature of scientific information, the fishery resources, and the fisheries.

Ecosystem Component Stocks

The 2009 guidelines suggest classifying fish stocks into two groups—stocks in the fishery and ecosystem component species. Stocks classified as being in the fishery would include certain target species and sometimes non-target species that the councils and/or the Secretary believe require conservation and management. To encourage ecosystem management, NMFS created the ecosystem component species group. Stocks in the fishery require determinations of their condition, reference points and ACLs while ecosystem component species do not. The guidelines define ecosystem component species or stocks as nontarget species, species not subject to overfishing, species not likely to become subject to overfishing or being overfished, and species not generally retained for sale or personal use. Although not considered to be in the fishery, the guidelines encourage councils to consider measures to protect the role of ecosystem component species in the ecosystem by minimizing bycatch and bycatch mortality.

Issues

Disagreement about the use of ACLs is related in part to management and scientific uncertainties. Regardless of whether stock assessments are uncertain, ACLs are required for all fisheries and in some cases ACLs may impose strict constraints on the fishery. To ensure ACLs are not exceeded managers are now taking precautionary approaches when determining allowable catch. These concerns with uncertainties and the effect on fisheries have prompted proposals to improve stock assessments or to exclude certain stocks from ACL requirements. Some regions such as the North Pacific, which has a history of using catch limits, have adjusted to ACL requirements, while fisheries which used indirect controls on harvest in regions such as the Northeast have been subject to greater controversy and social and economic disruption.[28]

Approaches to reducing uncertainty usually focus on technical improvements such as collecting more and better data and improving assessment models. Often many recommend dedicating more resources for data collection and stock assessments. They reason that by reducing risk of overfishing associated with uncertainty, the need for precautionary measures could be lessened. However, data and modeling improvements are likely to be costly and would require further increases of federal appropriations. The benefits of a closer approximation of benchmark population levels such as Maximum Sustainable Yield (MSY) are limited. Thus the value of these improvements must conform to the rule of diminishing returns.[29] As more resources are directed to this purpose the marginal benefits of lower uncertainty decrease. An unanswered question for many fisheries is whether the current management system is in need of greater investment or if it has already reached the level where costs of additional information are greater than the benefits derived from greater certainty.

"Data-Poor" Stocks

The causes of uncertainty vary by fishery and specific circumstances, but uncertainty plays a role in management decision-making for both data-poor and well-studied stocks. Data poor stocks are stocks for which there are inadequate data to complete a stock assessment to estimate biomass and fishing mortality reference points. In 2013, NMFS reviewed 478 individual stocks and stock complexes that are currently managed under 46 fishery management plans.[30] Of the 478, there were 178 with an unknown overfishing status and 258 with an unknown overfished status. Many of the data-poor stocks are of relatively low value or minor components of fisheries. However, it should be noted that many believe these stocks provide biological diversity and ecological value to the system.

One option open to managers is to use recent average catch as a basis for establishing ACLs. Another option is to group several stocks into a stock complex and use one or more indicator stocks within the complex. This option relies on the assumption that the stock complex can be managed and monitored using one or more stocks that can be assessed. Another option might involve moving stocks to the ecosystem components species category to exclude them from ACL requirements. Some have speculated that ACL requirements may work against conservation because it provides an incentive to designate stocks in the ecosystem category to avoid management.[31]

Well-Studied Stocks

Even those stocks which are relatively well studied are subject to management and scientific uncertainty because of data constraints, ecological factors, and mis-specified models. In 2013, according to NOAA, of the 478 stock and stock complexes that are currently managed under fishery management plans, there were 300 with a known overfishing status and 230 with a known overfished status. The Gulf of Maine cod stock is a recent example of the difficulties fisheries scientists face in assessing fish populations, even when the stock is relatively well-studied. Gulf of Maine cod is one of the most valuable species of the Northeast multispecies fishery and the mainstay of many inshore fishermen. The stock assessment reviewed at the Groundfish Assessment Review Meeting (GARM III) in 2008 indicated that overfishing was still taking place in 2007, but the stock was no longer overfished (stock biomass had increased above the level that defines it as overfished). Instead of further progress, the 2011 stock assessment reviewed at the 53rd Stock Assessment Workshop (53rd SAW) showed that overfishing continued in 2010 and stock spawning biomass was one-third of the level estimated in 2007 indicating the stock was also overfished. Moreover, updated information and the new population model showed that in 2007, the stock was actually overfished.[32]

Recreational Data and Uncertainty

Often a major source of management uncertainty is related to stocks that are taken in recreational fisheries. Recreational catch is difficult to quantify because landings are widely dispersed and taken by many different participants. When a significant percentage of catch is taken by recreational fishermen, it usually adds to uncertainties in developing stock assessments. Furthermore, recreational quotas are often difficult to manage on a real-time basis because of their nature. Overages may be common because catch is compiled using statistical models that may calculate totals months after the annual fishery is finished. These factors may lead to unpredictable recreational openings and closures and the use of accountability measures (AMs) in subsequent years that may limit quotas significantly.[33] Replacement of the Marine Recreational Fisheries Statistics Survey with the new Marine Recreational Information Program is focused on improving recreational data, but it may take several years before this information can be fully incorporated into the management process. Regardless, there will continue to be substantial uncertainty related to recreational harvests, especially when using recreational information to account for landings during the fishing year.

Bills Introduced During the 112th Congress

During the 112th Congress, H.R. 2304, H.R. 6350, and S. 1916 included provisions that would have excluded certain stocks from ACL requirements. These bills focused on the need for more timely stock assessments and would have stopped managers from establishing ACLs without periodic stock assessment updates. These bills also would have defined and applied the concept of an "ecosystem component stock" or "ecosystem component species" in statute and excluded them from ACL requirements.[34]

H.R. 3061 and H.R. 6350 would have amended Section 304 of the MSFCMA by adding a provision that would have allowed the Secretary to suspend ACLs. Suspension of ACLs would have been allowed if the Secretary determines that the fishery is not overfished or approaching a condition of being overfished, if any stock of fish in the fishery previously affected by overfishing is rebuilt, and if the Scientific and Statistical Committee (SSC) cannot ensure that the fishery management plan for the fishery is consistent with Section 301(a)(8).[35] Section 301(a)(8) requires that conservation and management measures provide for the sustained participation of fishing communities and to the extent practicable, minimize adverse economic impacts on such communities.

H.R. 3061 also included a section which would have required the Secretary to enter into an agreement with the National Research Council (NRC) to study current implementation of recreational survey methods. The study would have updated the assessment of recreational survey methods that NRC published in 2006. The study also would have evaluated the extent to which recommendations made in 2006 have been implemented and examine limitations of the Marine Recreational Information Program.

Limited Access Privilege Programs (Catch Shares)

Catch shares is the general term for fishery management systems which divide the total quota or harvest level of fish into individual shares or quotas. These shares may be allocated among different entities such as fishermen, cooperatives, or fishing communities. Other common terms for catch shares include individual transferable quotas (ITQs) and Individual Fishery Quotas (IFQs). The term limited access privilege (LAP) as defined in the MSFCMA is a specific type of catch share program which may be allocated to a person as opposed to a sector, cooperative, or fishing community.[36]

The term 'limited access privilege'—

(A) means a Federal permit, issued as part of a limited access system under section 303A to harvest a quantity of fish expressed by a unit or units representing a portion of the total allowable catch of the fishery that may be received or held for exclusive use by a person; and

(B) includes an individual fishing quota; but
(C) does not include community development quotas as described in section 305(i).

Catch share programs have been controversial since the first federal program was established for the Mid-Atlantic surf clam fishery in 1990. In the Sustainable Fisheries Act of 1996 (P.L. 104- 297) Congress responded to concerns related to the fairness of quota allocations, the potential for quota consolidation, and economic effects on the fishing industry and fishing communities by placing a moratorium on creating new ITQ programs.[37] Congress later extended the moratorium to September of 2002 after which it was allowed to expire. Since the moratorium expired, ten new programs have been established bringing the total number of federal catch share programs to fifteen.

Typically without catch shares, annual quotas or annual catch limits are established by fishery scientists and managers and made available to all fishermen who have permits to operate in the fishery. When the annual quota is reached, the fishery is usually closed to prevent overfishing of fish stocks. Management with a fishery-wide quota provides an incentive for fishermen to gain the greatest share of the total quota, often as quickly as possible, before the fishery is closed. This outcome has been characterized as "derby fishing" or the "race to fish."

Academic research and evaluations of existing catch share programs have shown that catch shares can change incentives and improve economic efficiency.[38] Allocating the total quota among permit holders changes incentives because entities such as individual fishermen or cooperatives possess a secure share of the quota. By providing fishermen with their own quota, investments in vessels, equipment, and crew to compete with others for a greater share of the total quota become unnecessary. Instead investments are likely to be more closely aligned with the individual fisherman's allocation. If quota shares are transferable, vessel owners may purchase or sell quota to match the needs of their business.

Catch shares may also provide fishermen with greater flexibility to land fish when conditions such as markets and weather are most favorable. Markets often improve under catch share systems because landings can be spread out over a longer period instead of shortened seasons with high landings and lower prices. Generally, as fishermen gain greater control over the resources that are allocated to them, such as when to fish, their individual businesses and the harvesting sector become more profitable.

Issues

Catch shares have remained controversial because of potentially higher management costs, the perceived fairness of the initial allocation, concerns with consolidation of the fleet and associated loss of employment, and effects on fishing communities. There also appears to be a general perception among some in the fishing industry that catch share programs have been imposed on the industry.

Catch shares may increase management costs because of administrative costs to set up and operate programs, and monitoring costs to ensure that individual quotas are not exceeded. Monitoring of catch at-sea by observers can be costly and burdensome, especially for smaller vessels. For some programs, costs are recovered by NMFS through fees and observers contracted by the fishing vessel or company. In other cases, the agency has implemented programs and shared management costs, but plans to phase out assistance. In these cases, especially where stock rebuilding is occurring, the share of costs between the fishing industry and government has become contentious.

The initial allocation of catch is especially controversial because the basis for allocating harvest among fishermen (e.g., historical participation, auctions, or others) will favor some fishermen over others. In some cases, those fishermen who are allocated quota gain a onetime windfall equal to the discounted value of all future profits from the individual quota. Some perceive this outcome as unfair to others in the industry such as crew members or future fishermen. Some fishermen who are not allocated enough quota to be economically viable, may have to sell their share to those with more capital. For those who want to leave the fishery this could be an advantage, but not for those who do not wish to leave. Catch shares may shift landings and affect specific fishing ports disproportionately. For example, in the Alaska halibut catch share (IFQ) fishery, small remote fishing communities have lost fishing rights because residents have been more likely to sell than buy quota.[39]

A related concern is that disproportionate shares of quota could be controlled by a relatively small number of fishermen. Some fishermen are concerned that this will change the nature of fisheries and make small scale fishing unprofitable. If consolidation occurs, investments in gear and vessels may be utilized more efficiently, but the number of crew employed in the fishery may also decrease. A possible consequence of greater fleet profitability is that remaining jobs may become more stable and permanent. Some have concluded that consolidation is inevitable in cases where fisheries have been overcapitalized, but that the redistribution of fishing rights may have unintended consequences.[40]

Previous Bills

Several similar bills were introduced in the 112th (H.R. 1646, H.R. 2772, H.R. 6350, and S. 1678) and 113th (S. 221) Congresses to address concerns related to LAP fisheries.[41] All five bills only would have applied to fisheries in the New England, Mid-Atlantic, South Atlantic, and Gulf of Mexico fishery management regions. In each case a petition requesting development of the LAP program would have been required and approval of the proposed LAP (catch share) program would have depended on a vote of eligible fishermen. H.R. 1646 would have terminated programs after five years unless two-thirds of eligible fishermen approved continuation of the program. S. 1678, H.R. 2772, and S. 221 would have terminated LAP programs if the Secretary determined that the number of eligible fishermen in the fishery decreased by 15% from the year before the program was established.[42] S. 1678, H.R. 2772, and S. 221 would have required fees to recover all costs of LAP programs including observer costs. H.R. 6350 also would have added the definition of the term catch share to the MSFCMA.

Management Process and Decision-Making

Fisheries managers are challenged to take management actions which both minimize uncertainty and incorporate uncertainty in the decision making process. Some have questioned whether NOAA has the data and science to properly manage fisheries under current overfishing and stock rebuilding requirements.[43] Management actions such as conservative harvest limits and fishery closures have been questioned when data are limited and stock assessments are perceived by many to be uncertain. In addition to improving data and requiring more timely stock assessments, some have proposed taking risk neutral approaches when estimating ACLs, broadening peer-review requirements, and constraining management decisions perceived by fishing interests to be extreme.[44]

Issues

According to NOAA guidelines, scientific and management uncertainty should be incorporated by setting ACLs according to precautionary or risk-averse approaches. Many fishermen are concerned with a risk-averse approach because they believe fisheries are often constrained unnecessarily. They assert that management should be risk neutral and management actions that would constrain the fishery should not impose abrupt and severe measures. Conversely, environmentalists have advocated for precautionary approaches because of historic tendencies of managers to take risks by using optimistic assumptions to set quotas. They also contend that uncertainty should not be used to undermine the best available scientific information or as an excuse for inaction.

Some are concerned that the interests of NMFS scientists sometime diverge from those of the fishing industry. They have advocated for more external peer review of stock assessments to ensure impartiality and to more fully consider different views. These changes also would be likely to convince some fishermen that the process is more fair and balanced. On the other hand, some would argue that the current process provides adequate peer review and that the best available science is currently used in the management process. They claim that further reviews would not add significantly to current assessments and that costs limit the amount of data and complexity of fishery models that might be used.

Concerns have been expressed by many in the fishing industry that the management process needs greater scrutiny when management measures harm fishing businesses. Management actions such as closures have been especially controversial in Gulf of Mexico and South Atlantic recreational fisheries. They assert the abrupt management actions harm businesses and make planning difficult. Fishermen also have claimed that these management actions are often based on uncertain and dated assessments. They surmise that the process needs to ensure that decisions are based on current conditions and the process is more open and accessible to those being regulated. Often fishermen cannot attend FMC meetings because they are at-sea fishing, especially when fishing or market conditions are favorable.

Bills Introduced During the 112th Congress

Several bills would have changed RFMC management processes and decision making. H.R. 1646, would have directed RFMC scientific and statistical committees to provide "risk neutral" scientific advice to the FMCs and limit SSC recommendations to change ACLs unless the basis of these recommendations are peer reviewed by non-governmental entities.[45] H.R. 1646 also would have limited closures by requiring certain conditions are met before a fishery could be closed and require review of the effects of the closures on small businesses and jobs in coastal communities. H.R. 1646 and

H.R. 6350 also would have required the Secretary to report to Congress on the effects and characteristics of fishery closures established during the previous five years.

Often fishermen are unable to attend RFMC meetings because they are at sea. H.R. 2753 and H.R. 6350 would have required each RFMC to make live broadcasts of RFMC, Scientific and Statistical Committee, and the Council Coordination Committee meetings available on the Internet to allow for greater levels of public participation. RFMCs also would have been required to provide complete audio, complete video, and transcripts of certain meetings depending on the circumstances.

H.R. 3061 included a section related to reports of the SSCs. The SSCs would have been required to provide an annual report on the process and information used in providing scientific advice to its Council. Each Council also would have been required to submit to the Secretary and make available to the public any reports or other information provided by the SSC.

Ecosystem Management

Many advocates, managers, and scientists support managing fisheries at the ecosystem level because of its potential to include factors often not included in fishery assessments. Most fishery management actions still depend on single species stock assessments. Current stock assessments focus on the relationship between fishing mortality and fish stocks, but fish populations are also affected by other elements of the ecosystem such as predators and prey, competition, environmental conditions, and other factors. Supporters of ecosystem-based management stress that when making management decisions ecosystem-based data and models are needed to incorporate the true complexity of the marine environment. Despite general agreement on the relevance of ecosystem management, questions remain with regard to its implementation and cost.

In 1996, Section 406 of the MSFCMA required the Secretary of Commerce to establish an advisory panel and for the panel to report with recommendations to expand the application of ecosystem principles in fisheries management. In 1998, the panel completed its report which assessed the extent to which ecosystem principles were applied in fishery research and management and how to further integrate ecosystem principles into future fishery management and research. The panel noted that

> a comprehensive ecosystem-based fishery management approach would require managers to consider all interactions that a target fish stock has with predators, competitors, and prey species; the effects of weather and climate on fishery biology and ecology; the complex interactions between fishes and their habitat; and the effects of fishing on fish stocks and their habitat.[46]

In their report, the panel described the difficult task of managing at the ecosystem level and recognized that, in most cases, available data were insufficient. However, it stressed that there are practical ways to use the information that is available and recommended the use of fishery ecosystem plans (FEPs) to further incorporate ecosystem principles into FMPs. The FEP would document the structure and function of the ecosystem in which fishing activities occur as well as provide information to managers about the effects of their decisions on other components of the ecosystem and the effects of other ecosystem components on fisheries. The panel concluded that

> if fishery management is to further incorporate ecosystem principles, Congress must provide a specific mandate to NMFS and the regional councils to do so and must fund the scientific infrastructure required to support the decision-making process. Requiring regional councils to prepare FEPs provides a mechanism to focus and inform fishery management, to measure progress toward implementation of ecosystem-based fishery management, to identify research needs and ultimately to insure healthy and productive ecosystems.

In 2006, Section 406 of the reauthorized MSFCMA required NMFS to undertake a study of the state of the science for advancing the concepts and integration of ecosystem considerations in federal fisheries management.[47] Section 406 specified four objectives including the following:

(1) form recommendations for scientific data, information, and technology requirements for understanding ecosystem processes and methods for integrating this information from federal, state, and regional sources;
(2) form recommendations for processes for incorporating broad stakeholder participation;
(3) form recommendations for processes to account for effects of environmental variation on fish stocks and fisheries; and
(4) describe existing and developing RFMC efforts to implement ecosystem approaches.

The report recommended maintaining and expanding current fishery-dependent and fishery-independent surveys, and collection of data for long-term studies. It also supported research to improve ecological models that are needed to improve understanding of dynamic ecosystem processes. Investigation of the effects of environmental variation on fish stocks, especially climate change, was emphasized by the report. Generally the report acknowledged that there are still critical gaps in the data and modelling of marine ecosystems.

Issues

Despite progress, data requirements of ecosystem-based management are extensive and assessments and implementation would be extremely complex. According the 2009 NOAA report:

> The ecosystem approach to management (EAM) is a more complex and information intensive than traditional fisheries management approaches and will require dedicated resources to implement effectively. At present, Councils cannot undertake EAM as a dedicated programmatic task and NMFS is unable to provide the required environmental and fisheries data and associated predictive analyses.

The report stated that the present ability of existing stock assessments to account for environmental effects is minimal, current multispecies models have very limited predictive accuracy, and ecosystem shifts can generally be recognized only after they have occurred. Most current surveys do not provide enough information to manage all stocks or to provide sufficient understanding of the relationship among habitat, benthic organisms and fish species. Moreover, management of many ecosystems components which affect the productivity and abundance of fish populations such as water quality or wetlands are outside the authority of fishery managers.

The eight RFMCs have taken different paths and have exhibited different amounts of progress toward incorporating ecosystem principles into fisheries management. Generally there is a lack of agreement among RFCMs of how to implement ecosystem approaches to fisheries management.[48] For example, some RFMCs have developed FEPs while others have not. Unresolved questions include the levels of investment needed to achieve specific ecosystem management objectives and whether specific statutory changes are needed to implement ecosystem management.

At a recent forum, three inter-related ecosystem topics were introduced to bring greater focus to implementing ecosystem-based management. Some reason that it may be more practical to take an incremental approach by concentrating on more specific and immediate concerns. Focus areas included the following:[49]

- adapting to climate change;
- managing forage fish; and
- integrating habitat considerations.

Climate Change

Climate change is likely to cause shifts of ecosystems and affect the composition and productivity of related fish stocks. Some have questioned how fisheries managers can prepare and mitigate for these shifts and changes in productivity.[50] As fish distributions shift it is likely management will require greater coordination among national (adjacent RFMCs) and international jurisdictions. As systems are modified by climate change managers may need to take approaches that are more proactive and precautionary. For example, very little is known about Arctic fish stocks that are becoming more accessible to fisheries. The Pacific RFMC adopted and the Secretary of Commerce approved the Arctic FMP which closed U.S. federal waters of the Arctic Ocean to commercial fishing until sufficient data has been collected to guide management and exploitation.

Forage Fish

Forage fish such as herring and anchovies play an important role in marine ecosystems. They comprise a significant portion of total ecosystem biomass and they are consumed by predators throughout their life span.[51] Forage fish provide an important link between primary production (phytoplankton) and higher tropic levels (predators such as tuna). Commercial fishermen harvest forage fish for both direct consumption and indirect uses such as bait, fishmeal, and fish oil. Some recreational and environmental groups argue that greater protection for forage fish is needed because of their role in the ecosystem. They question whether the characteristics of forage fish warrant unique management approaches and if so, they ask whether the RFMCs have the flexibility to address these concerns under current law and regulations.

Essential Fish Habitat

The 1996 reauthorization established requirements to identify, describe, conserve, and enhance essential fish habitat (EFH). The MSFCMA defines EFH as "those waters and substrate necessary to fish for spawning, breeding, feeding or growth to maturity."[52] EFH is now identified and described for species under each FMP and in some cases specific areas have been closed to fishing because of the impacts of fishing gear on bottom habitat. The distinction between essential and non-essential habitat, however, is problematic because in some cases regional councils have not distinguished between *essential* and *non-essential* parts of a range of habitat,

choosing instead to define all habitat as essential. Prioritizing among habitats is needed if managers are to achieve measurable progress and focus is critical in this age of shrinking budgets.[53] In most cases, current understanding of the linkages between habitat and fish productivity and the extent of harm to habitat caused by fishing and non-fishing activities is either insufficient or unavailable.[54]

Nonfishing impacts on habitat fall outside the authority of fishery managers although many estuaries and other nearshore fish habitats are threatened by nonfishing activities. The MSFCMA requires federal agencies to consult with the Secretary of Commerce when their actions may adversely affect essential fish habitat.[55] Some question how consultation on nonfishing habitat degradation can be improved and whether legislative action is needed to minimize nonfishing impacts.

Bycatch

The selectivity of commercial fishing gear depends on its characteristics and the nature of the species that it targets. Sometimes fishermen cannot control for the size of fish or species that are caught in the course of fishing. National standard 9 of the MSFCMA requires conservation and management measures to minimize bycatch and bycatch mortality.[56] Bycatch is defined in the act as fish harvested in a fishery, but not sold or kept for personal use, and includes economic discards and regulatory discards.[57] Economic discards are fish that are targeted by the fishery, but are not retained because they are of an undesirable species, poor quality, or for other economic reasons.[58] Regulatory discards are fish harvested in a fishery that fishermen are required by regulation to discard whenever caught, or to retain but not sell.[59]

The MSFCMA's definition of bycatch explicitly excludes fish released alive under a recreational catch-and-release fishery management program. The MSFCMA definition does not include incidentally caught sea turtles, sea birds, and other non-fish organisms. For its national strategy, NOAA defines bycatch more broadly as discarded catch of any living marine resource, plus unobserved mortality due to a direct encounter with fishing gear.

The MSFCMA also includes several sections that focus on reducing bycatch. Section 303(a)(11) requires FMPs to establish a standardized reporting methodology to assess the amount and type of bycatch occurring in the fishery, and include conservation and management measures that, to the extent practicable and in the following priority—

(A) minimize bycatch; and
(B) minimize the mortality of bycatch which cannot be avoided;

The act also includes a section concerning North Pacific Fisheries Conservation that focuses on bycatch reduction incentives and a section which established a bycatch reduction engineering program.[60]

Since most bycatch is discarded at sea, for many fisheries it has been difficult to quantify its extent and composition. In 2011, NOAA released a national bycatch report that provided bycatch estimates for federal commercial fisheries and recommendations for improvements to bycatch data collection and estimation.[61] Bycatch data are provided by onboard observers, self-reported vessel logbooks or trip reports, commercial dealer landings reports, and protected species reports. The report developed the following tier system to rate the 152 fisheries included in the report.

- Tier 0 - 24% of fisheries reviewed. No estimates exist for these fisheries because of the lack of data.
- Tier 1 – 16% of fisheries reviewed. Bycatch data were available but generally unreliable or data had not been analyzed.
- Tier 2 – 15% of fisheries reviewed. Bycatch estimates were generally available but would have benefited from improvement in data or analytical methods.
- Tier 3 – 41% of fisheries reviewed. Bycatch estimates were generally available and data were of higher quality than Tier 2.
- Tier 4 – 4% of fisheries reviewed. Bycatch estimates were available and based on the highest quality data and analytical methods.

The report estimated that in 2005 national bycatch totaled 1.221 billion pounds while landings of these fisheries totaled 6.068 billion pounds. NMFS intends to use this first national compilation of bycatch as a baseline for subsequent updates. The report notes that data were from 2005 and that it is likely that the quality of data and estimates have improved since then, and that further bycatch reductions have been made in some fisheries.

Issues

Despite efforts to improve bycatch data collection and to address the need to reduce bycatch, these issues are still of concern to the public, environmental organizations, recreational anglers, and commercial fishermen. Bycatch can harm marine ecosystems, deplete protected species (endangered species and marine mammals), and reduce marine biodiversity.[62] Economic losses also occur when undersized fish are discarded after dying in fishing gear (regulatory discards) and valuable fish are caught in a fishery that is not allowed to retain them (salmon caught in the Alaska pollock fishery).

RFMC and fishing industry efforts have decreased bycatch in many fisheries. For example, an innovative industry program in the North Pacific avoids bycatch "hotspots" by closing areas of the pollock fishing grounds when Chinook salmon bycatch rates are high in those areas. The North Pacific RFMC has also established bycatch limits, closed areas, and gear requirements for trawl fisheries operating in the region.[63] However, according to Oceana, an environmental group, only 20 percent of existing FMPs include incentives to minimize bycatch.[64] Many, especially environmental groups, support efforts to further establish bycatch reduction incentives, set bycatch caps, and develop more selective fishing gear.

Adequate knowledge of the quantity of organisms discarded is needed to measure fishing mortality and to develop reliable stock assessments. For example, accounting for by-catch during development of ACLs is required, but often managers are hampered by the paucity of bycatch data. According to some environmental groups, bycatch information is unreliable and inconsistent because of insufficient at-sea coverage by observers or electronic monitoring. Although observers are stationed on fishing boats, the level of coverage is often below levels needed to accurately quantify bycatch. Observer coverage is costly and in some cases observers are difficult to accommodate on small vessels. Electronic monitoring systems are currently under development in several regions. In some cases these systems may supplement direct observations and offer a less costly alternative.

Other Issues

Fishery Disaster Assistance

Disaster relief may be provided by the federal government to assist the fishing industry when it is affected by a commercial fishery failure. A commercial fishery failure can be declared when fishermen endure economic hardships resulting from fish population declines or other disruptions to the fishery. The Department of Commerce can provide disaster assistance under Sections 308(b) and 308(d) of the Interjurisdictional Fisheries Act (IFA; 16 U.S.C. §4107), as amended, and Sections 312(a) and 315 of the MSFCMA (16 U.S.C §1861). NMFS determines whether a commercial fishery failure has occurred and in allocating federal funds to states and affected fishing communities. Congress plays a pivotal role by appropriating funds and providing oversight of the process. States also play a central role by initiating requests, providing information, planning for the use of funds, and often disbursing funds.

Critics contend that disaster assistance programs often fall short of expectations because sometimes funds are not disbursed in a timely manner. There is no permanent relief fund to draw from when fishery failures occur and funds are seldom appropriated in anticipation of disasters. Given the timing of appropriations bills and congressional schedules, it can be difficult to appropriate funding in a timely manner. Furthermore, in several cases it has taken over a year for the Secretary of Commerce to make a determination.[65]

During the 112th Congress two bills (H.R. 1646 and H.R. 6350) included provisions which would have required the Secretary to make a determination within 60 days of the date on which the Secretary received the request. Two provisions were included in the Agriculture Reform, Food, and Jobs Act of 2013 (S. 954) which passed the Senate. The first would have required the Federal Crop Insurance Corporation (FCIC) to develop a feasibility study to determine the best method of insuring harvesters. The FCIC would be required to submit a report with the results of the study to the House Committee on Agriculture and the Senate Committee on Agriculture, Nutrition, and Forestry. The second provision would have added commercial fishermen to the list of eligible borrowers for emergency loans. The USDA's Farm Service Agency provides emergency loans due to drought, flooding, other natural disasters or quarantine. The House bill did not include similar provisions and neither of the Senate provisions was included in the final bill (P.L. 113-79).

Data Collection and Confidentiality

The need to monitor compliance and collect data, especially from vessels at-sea has increased the use of at-sea observers and hastened the development of electronic monitoring systems. Observers are often required on commercial fishing boats to monitor compliance, document take of protected species, and record biological data. The expense of observers and need for observation on small boats with limited space has encouraged the use of electronic monitoring. Electronic monitoring utilizes cameras and other electronic monitoring equipment to carry out some of the same monitoring tasks as observers. With the increase in at-sea monitoring new concerns with confidentiality and use of data have emerged.

Information collection is essential for developing stock assessments, developing regulations, and managing fisheries. Currently, information submitted to the Secretary, state agency, or fishery commission by any person in compliance with the MSFCMA is confidential and cannot be disclosed.[66] Under the MSFCMA and current confidentiality rules, when data are used for management purposes, it may only be disclosed when it is aggregated as a summary. Summary formats are used so that information of a specific business is not directly or indirectly disclosed. On May 23, 2012, NMFS released a proposed rule to implement the MSFCMA confidentiality provisions and to formalize current data confidentiality practices.[67]

Subsistence

Subsistence fishing is a significant activity for many who live in coastal areas and a source of resources for some communities, but it is not explicitly addressed in the MSFCMA. Subsistence fishing can satisfy diverse needs such as personal consumption and community traditions. Although subsistence fishermen benefit from management that sustains stocks, they can also be affected by allocation decisions and other management measures. Their needs, motivations, and use of fishery resources are likely to differ from those of most recreational and commercial fishermen. Subsistence fishing is not defined in the MSFCMA, but an example of a definition for subsistence uses in Department of the Interior regulations is[68]

> the customary and traditional uses by rural Alaska residents of wild, renewable resources for direct personal or family consumption as food, shelter, fuel, clothing, tools, or transportation; for the making and selling of handicraft articles out of nonedible byproducts of fish and wildlife resources taken for personal or family consumption; for

barter, or sharing for personal or family consumption; and for customary trade.

Some groups are concerned that fisheries regulations do not necessarily account for traditional values and cultural beliefs.[69] Some have questioned whether subsistence fishing could benefit by being explicitly included in the management process for decisions related to allocation of resources and disaster relief. For example, the halibut IFQ program appears to have led to the loss of fisheries participation in some small Alaska indigenous communities.[70]

International Agreements and Port State Measures

Coordinated management of fish stocks that travel among national zones of jurisdiction and the high seas is accomplished by international agreements which may establish regional fisheries management organizations (RFMOs). Each nation is expected to develop domestic laws and regulations that are consistent with each agreement. In many cases this requires implementing legislation and regulations to meet commitments to RFMOs and international fisheries agreements. Often legislation concerning RFMOs and other international fisheries issues is passed as part of MSFCMA reauthorizations. The following are two examples of legislation related to international fisheries management that have been introduced during the 113th Congress.

Illegal, Unreported and Unregulated Fishing and Other Fisheries Agreements

The Magnuson Stevens Fishery Conservation and Management Reauthorization Act of 2006 (P.L. 109-479) amended the High Seas Driftnet Fishing Moratorium Protection Act (P.L. 104-43) to address illegal, unreported, and unregulated (IUU) fishing. The Moratorium Protection Act requires the NMFS to identify nations engaged in IUU fishing; to consult with identified nations; and to determine whether the nation has taken actions to address IUU activity.[71] If the nation in question does not take actions to stop its IUU activities, U.S. imports of fisheries products from that nation may be prohibited.

During the 113th Congress, two similar bills have been introduced in the House (H.R. 69) and the Senate (S. 269) to harmonize and strengthen the enforcement provisions that implement international fishery agreements to which the United States is a party. Provisions of these bills focus on reducing IUU fishing activities. Both bills also would implement the Convention for the Strengthening of the Inter-American Tropical Tuna Commission established by the 1949 Convention between the United States of America and the Republic of Costa Rica (Antigua Convention). On November 18, 2005, the Senate ratified the Antigua Convention. On December 17, 2013, S. 269 was reported by the Committee on Commerce, Science, and Transportation, but no further action has been taken on either bill.

Agreement on Port State Measures

The Agreement on Port State Measures to Prevent, Deter and Eliminate Illegal, Unreported and Unregulated Fishing was approved by the United Nations Food and Agriculture Organization (FAO) Conference at its Thirty-sixth Session on November 22, 2009, through Resolution No. 12/2009, under Article XIV, paragraph 1 of the FAO Constitution. This Agreement, comprised of 37 Articles and 5 Annexes, aims to prevent illegally caught fish from entering international markets through ports. Under the terms of the treaty, foreign vessels will provide advance notice and request permission for port entry, countries will conduct regular inspections in accordance with universal minimum standards, offending vessels will be denied use of port or certain port services, and information sharing networks will be created. The United States signed the Agreement on November 22, 2009, the President transmitted it to the Senate on November 14, 2010, and on April 3, 2014, the Senate ratified the treaty.

In the 113th Congress, on February 11, 2012, the Pirate Fishing Elimination Act (S. 267) was introduced.[72] S. 267 would implement the agreement on port state measures and apply to vessels seeking entry to ports subject to the jurisdiction of the United States. The bill includes provisions related to duties of the Secretary, procedures for vessels entering U.S. ports, denial of port services, inspections, prohibited acts, and enforcement. On January 8, 2014, S. 267 was reported by the Senate Committee on Commerce Science and Transportation, but no further action has been taken.

State-Federal Relationship

Prior to 1976, states had management authority over all fisheries in waters adjacent to their coastlines, and there was little or no federal jurisdiction over living marine resources in these waters. With enactment of the Fishery Conservation and Management Act[73] in 1976, marine fishery resources within 3 to 200 nautical miles of shore came under federal jurisdiction, while states retained jurisdiction of marine fishery resources from their coastline out to three nautical miles offshore. In Section 306, the MSFCMA states that the act neither extends nor diminishes the authority or jurisdiction of any state within its boundaries. The MSFCMA attempts to balance state interests with federal conservation and management goals, through the composition of RFMCs, support of regional commissions, and coordination with state fishery management. Many species managed by the RFMCs are found and fished in both state and federal waters.

Some recreational and commercial fishermen are in favor of shifting greater management responsibility to state or regional levels because they believe local individuals are more knowledgeable about specific environmental conditions and economic needs of the state. These changes are likely to require more state resources to collect data, assess stocks, and develop management measures. In contrast, while recognizing the need for local expertise, some interests would oppose further decentralization of federal fisheries. Decentralized management, in their opinion, might put short-term local economic needs above long-term stock rebuilding and productivity.

The red snapper fisheries of the Gulf of Mexico are among the most controversial fisheries under federal management and some would like to see greater state involvement in its management. The Gulf of Mexico stock was fished to low levels during the 1990s, and the Gulf of Mexico RFMC developed several FMP amendments to rebuild the fishery. Quotas have been kept at low levels and although the stock is overfished (low biomass), overfishing is no longer occurring. As rebuilding has proceeded and stock biomass has increased, associated fishery quotas, catch rates, and the average size of fish have also increased. However, seasons have become shorter because these increases have allowed recreation fishermen to reach the red snapper quota more quickly. Fishermen question whether stock assessments accurately reflect red snapper abundance when they are observing high catch rates. Recently, the fishery has become further complicated by inconsistent seasons in state and federal waters. Some have advocated for a greater state role in management such as determining seasons and extending state jurisdiction over red snapper.

During the 113th Congress six bills (H.R. 1430, H.R. 3099, H.R. 3197, S. 681, S. 747, and S. 1161) have been introduced to transfer management authority over red snapper from federal to State or regional control. H.R. 1430, S. 681, and S. 747 would grant management authority to the coastal States in the region.[74] The authority would be contingent on agreement by the Governors on a red snapper fishery management plan. If the Governors fail to reach an agreement, management authority would revert back to the Secretary of Commerce. H.R. 3099, H.R. 3197, and S. 1161 would transfer authority over red snapper to the Gulf States Marine Fisheries Commission.[75] The

Commission would develop a fishery management plan that would take the place of the current Gulf of Mexico RFMC FMP. H.R. 3099 and H.R. 3197 would also require review and certification by the Secretary of Commerce that the plan is compatible with Section 301 of the MSFCMA and would ensure the long-term sustainability of the Gulf of Mexico red snapper population.

Use of Funding from Enforcement Penalties

NOAA's Asset Forfeiture Fund (AFF) is funded from fines collected from fishermen who violate marine resource laws. The MSFCMA provides the National Marine Fisheries Service (NMFS) Office of Law Enforcement (OLE) and NOAA's Office of General Counsel for Enforcement and Litigation (GCEL) authority to use funds for supporting fisheries enforcement activities.[76] In 2010, NOAA was the subject of an investigation into the potential misuse of the AFF.[77] The audit commissioned by the U.S. Department of Commerce, Office of Inspector General and conducted by an outside accounting firm found that OLE's and GCEL's management processes and internal controls were nonexistent or weak. These problems were reflected in OLE's procurement and management of vehicles and vessels, OLE's and GCEL's use of the AFF for international travel, and OLE's administration of its Special Operations Fund for covert and undercover operations. In 2010, NOAA began implementing reforms to policies and procedures in response to the Inspector General's report.[78]

In response to concerns related to the use of funds and needs for funding of fisheries research and monitoring, five bills were introduced during the 112th Congress that addressed concerns with NOAA enforcement funds. Four of these bills (H.R. 1013, H.R. 2610, H.R. 6350, and S. 1312) proposed to use fines, penalties, and forfeitures of property for fisheries management needs such as monitoring and data collection. For certain fisheries enforcement cases, three bills, H.R. 2610, S. 1304, and S. 1312 would have used funds to reimburse fishermen for legal fees and costs incurred from fisheries enforcement penalties.

Aquaculture

Development of commercial aquaculture facilities in federal waters is hampered by an unclear regulatory process in the EEZ, and technical uncertainties related to working in offshore areas. Regulatory uncertainty has been identified by the Administration as the main barrier to developing open ocean aquaculture. Uncertainties often translate into barriers to commercial investment. Controversy related to potential environmental and economic impacts have also contributed to slowing expansion. Legislation was introduced during the last several Congresses to establish a regulatory system for offshore aquaculture in the U.S. EEZ. Bills have also been introduced to prohibit aquaculture in the EEZ unless a law is passed to authorize such actions.

Nonetheless, on January 28, 2009, the Gulf of Mexico RFMC voted to approve a plan to issue aquaculture permits and regulate aquaculture in federal waters of the Gulf of Mexico. On September 3, 2009, the plan took effect because the Secretary of Commerce declined to oppose it within the required statutory period. Draft regulations to implement the plan are still under review at NOAA and according to NOAA could be completed sometime in 2014.[79] Some environmental groups have questioned whether NOAA has the authority to regulate aquaculture under the MSFCMA. NOAA's position is explained in a memorandum concerning its authority to regulate aquaculture with the following passage.[80]

> The Magnuson-Stevens Act does not expressly address whether aquaculture falls within the purview of the Act. However, the Magnuson-Stevens Act's assertion of exclusive fishery management authority over all fish within the EEZ, its direction to fishery management councils to prepare fishery management plans for any "fishery" needing conservation and management, together with the statutory definitions of "fishery" and "fishing," provide a sound basis for interpreting the Act as providing authority to regulate aquaculture in the EEZ.

It is likely that the controversy concerning aquaculture development and regulation in federal waters will continue. It remains an open question as to whether aquaculture can or should be regulated under the MSFCMA and whether legislation that addresses aquaculture development and management is warranted.

Seafood Certification

Ecolabels and seafood guides are among the most well-known programs that inform the choices of consumers and seafood buyers. The main goal of most seafood programs is to promote conservation and management by identifying seafood produced by sustainable fisheries.[81] Groups that provide or support ecolabels and seafood guides contend that consumers will choose seafood certified as produced by sustainable methods over seafood that is not certified. Producers of seafood may also obtain a price premium if consumers are willing to pay extra for an environmentally friendly product. If consumers are willing to pay the price premium, the ecolabel could provide an economic incentive for fisheries to adopt sustainable fishing practices.[82]

Ecolabeled seafood is recognized by a seal that is placed on the product's packaging to certify that it has met specific criteria related to its production. Criteria are developed by a standard setting organization such as the Marine Stewardship Council (MSC). These criteria are used to evaluate the fishery and determine whether it can be certified as sustainable. Certification of the chain of custody is also needed to ensure that seafood from the certified fishery is kept separate from uncertified seafood. Finally, wholesalers, processors, and retailers who sell certified seafood are required to follow certain standards regarding use of the ecolabel.

Another way to convey information related to seafood sustainability is through published guides. Seafood guides range from pocket guides used by consumers in restaurants and grocery stores to more comprehensive online guides. Guides consist of seafood product lists that are ranked by categories such as best choice, good alternative, and avoid.[83] Some lists identify general categories such as species or the common name of a seafood product, while others also identify specific fishing gear used to catch the product or harvest locations of the fishery. However, information available in supermarkets or restaurants is usually too limited to allow consumers to determine the specific fishery or capture method of a given product.

Some have questioned who should certify seafood products and which criteria are appropriate. According to some nongovernmental organizations independent seafood certification programs are needed because governments have failed to adequately manage and conserve fisheries resources.[84] Seafood certification is regarded as an alternative to relying solely on government action and the absence of a U.S. government label for seafood has left a gap which private companies and non-governmental organizations are trying to fill.[85] Some have countered that in the United States federal and state governments have been successful in arresting the decline of fishery resources. They question the need to certify fisheries that are already subject to some of the strictest regulatory requirements in the world. Some contend that when implemented and enforced, the 10 MSFCMA national standards may be sufficient to ensure sustainability.[86]

It is often difficult to find agreement on the definition of sustainable fisheries and on standards that should be used to measure sustainability. Seafood eco-labels and guides may also focus on a variety of different concerns such as ecosystem impacts, aquaculture methods, and health issues.

Differences in evaluation criteria have resulted in discrepancies among recommendations promoted by different groups. Some observers have recommended that some degree of government participation in ecolabeling programs could ensure the veracity of labels[87] and provide a recognizable standard for consumers.

Safety at Sea

Commercial fishing is one of the most dangerous occupations in the United States. From 2000- 2010, on average 46 fatalities occurred each year on fishing vessels, a rate of 124 deaths per 100,000 workers. The average rate over the same time period for all U.S. workers was 4 deaths per 100,000 workers.[88] Under the Commercial Fishing Safety Act of 1988, the U.S. Coast Guard is responsible for developing regulations for safety equipment and vessel operating procedures. The National Institute for Occupational Safety and Health collects data related to fatal injuries, establishes why fatalities occur, and designs and implements interventions. These efforts have reduced fatalities from an average of over 100 deaths per year prior to 1988.[89]

Some commercial fishermen contend that it is imperative for fishery managers to explicitly consider whether or not fishery management regulations will compel fishing captains and crew to work under unsafe conditions. National standard 10 (16 U.S.C. 1851(a)(10)) of the MSFCMA requires that conservation and management measures promote the safety of human life at sea. However, some question whether this standard should be revised to make safety considerations a priority when developing plans to manage fisheries.[90] They recommend that RFMCs should be required to address whether safety of fishermen is affected by management measures that encourage vessels to stay out in poor weather, work farther offshore, limit crew or vessel size, or make other changes that could increase risk.

The MSFCMA reauthorization in 2006 added Section 303(a)(9)(C) which states that fishery impact statements shall access the effects of conservation and management measures on the safety of human life at sea, including whether and to what extent such measures may affect the safety of participants in the fishery. NOAA has been developing an update to guidelines published in 1998 to address this provision.[91] On April 21, 2011 a proposed rule was published in the *Federal Register*, but no further action has been taken.[92]

Relationship to Other Selected Laws

National Environmental Policy Act

When the MSFCMA was reauthorized in 2006, Section 304(i) required NMFS to revise and update procedures to comply with the National Environmental Policy Act (NEPA; 42 U.S. C. 4231 et seq.) for actions related to fisheries management. Issues were related to mismatches in the timing of NEPA and MSFCMA processes and roles and responsibilities of NMFS and RFMCs. In response to the MSFCMA requirement, NMFS consulted with the Council on Environmental Quality (CEQ) and RFMCs, and sponsored public meetings. It then proposed a rule to better align RFMC and NEPA analytical and procedural requirements. The final draft rule was submitted by NMFS to the Office of Management and Budget (OMB) and at OMB's request withdrawn. Instead NMFS developed an Interim Policy Directive that focuses on roles and responsibilities, coordination of NEPA and MSA procedures such as timing, issues pertaining to NEPA documentation, and partnerships and efficiencies.[93]

According to NMFS, the policy directive satisfies the requirements of the NEPA provision. However, the RFMCs questioned whether the policy directive satisfies the intent of the MSFCMA to provide a more timely alignment of MSA and NEPA processes and a more streamlined environmental review process.[94] Furthermore, the RFMCs claimed that consultation during development of the policy directive was inadequate.

Marine Mammal Protection Act

The Marine Mammal Protection Act (MMPA, 16 U.S.C. §§1361 et seq.) established a moratorium on the taking of marine mammals in U.S. waters or on the high seas. Some types of fishing gear can accidentally take marine mammals during commercial fishing operations. Exceptions to the moratorium are provided through permits to take marine mammals during the course of commercial fishing and other activities. Under the MMPA the Secretary of Commerce acting through NOAA Fisheries is responsible for the conservation and management of whales, dolphins, porpoises, seals, and sea lions while the Secretary of the Interior acting through the Fish and Wildlife Service is responsible for walruses, sea otters, polar bears, manatees, and dugongs. Ongoing issues are often related to the lack of data and associated uncertainties in assessing human-caused mortality and the status of marine mammal stocks. Another issue involves the concern that the some marine mammal populations consume a growing proportion of potential fisheries yield.

National Marine Sanctuaries Act and Antiquities Act

The National Marine Sanctuaries Act (NMSA; 16 U.S.C. §§1431, et seq.) authorizes NOAA to designate specific sites for comprehensive and coordinated management and conservation. The broad NMSA mandate allows NOAA to designate areas to preserve or restore conservation, ecological, aesthetic, or recreational values of the designated areas. It requires the development and implementation of management plans, which serve as the basis for prohibiting or limiting incompatible activities.[95]NOAA has designated 13 sanctuaries, ranging in size from less than a square nautical mile to more than 100,000 square miles. Each site was designated for a specific reason, ranging from protecting cultural artifacts to protecting entire ecosystems. Since the management plans and regulations have been developed individually for each sanctuary and each sanctuary was established for a specified reason, they vary widely in how uses are managed and what uses are permitted.

Fisheries are not regulated in most sanctuaries and fisheries conservation is not the primary objective of any of the 13 national marine sanctuaries. In considering whether to regulate fishing, the NMSA provides the appropriate RFMC the opportunity to determine whether sanctuary fishing regulations are needed and to draft fishing regulations. The Secretary of Commerce must accept the Council's proposals or determinations unless they fail to fulfill the purpose of the Sanctuary's designation. In this case NOAA could prepare regulations under either the MSA or the NMSA.[96]

The Antiquities Act of 1906 (16 U.S.C. §§431-443) allows the President to proclaim locations of scientific or historical interest as national monuments and has been used for several marine areas.[97] Some have asserted that when the 1906 Antiquities Act was passed, Congress did not envision it use for vast marine areas.[98] They question whether recent presidential proclamations to establish marine national monuments are an abuse of executive power. Use of the Antiquities Act has raised objections that proclamations do not provide the opportunity for public input and debate.[99] Others counter that the act should be used to designate protected areas in the marine environment because it can be used expeditiously. Regardless, applying the Antiquities Act to marine areas will still require "negotiation, education, and consensus-building" including congressional funding commitments and involvement of local committees representing interested and affected parties.[100]

Endangered Species Act

The Endangered Species Act (ESA; P.L. 93-205, 16 U.S.C. §§1531-1543) provides for protection, and recovery of species of animals and plants that are threatened with extinction. Under the ESA, species can be listed as endangered or threatened according to assessments of their risk of extinction. Once a species is listed, legal tools are available to aid its recovery and protect its habitat. These tools may restrict activities that affect the listed species. Fishing may interact with protected marine species and in some cases fishing activities may be constrained to minimize harm to listed species.

HOUSE AND SENATE COMMITTEE PROPOSALS

The House Committee on Natural Resources and the Senate Committee on Commerce, Science, and Transportation have been pursuing efforts to reauthorize the MSFCMA during the 113th Congress by holding hearings and by releasing discussion drafts composed of potential amendments to the act. The House draft has been posted on the Natural Resources Committee website to gather public input on proposed changes. Proposals include topics related to flexibility in rebuilding fish stocks and setting ACLs, standards and procedures for developing catch share programs, relationships to other laws, data confidentiality, and red snapper management. The Senate Committee draft has been distributed to different stakeholders and has been posted on several websites. The draft addresses topics such as subsistence uses, forage fish, use of capital funds, rebuilding timeframes, and review of allocation among sectors. Both Committees intend to release refined proposals and continue progress on reauthorization issues during 2014.

CONCLUSION

Although science is often looked to for answers, societal values also play an important role when developing national policies. Often the fishing industry faces hardships associated with the natural variability of marine fish populations and management efforts to sustain populations above specific levels. These hardships are compounded when stock rebuilding becomes necessary. A critical fisheries policy question is whether it is in the national interest to provide greater management flexibility, increase resources for management and research, and generally expand support for commercial fishing and associated communities.

Seafood is the last major food source that depends on harvest of wild populations from the natural environment. When fisheries were being developed fishermen often had flexibility to target different species or move to different areas if resource abundance decreased. As limits to natural production were reached and most fish stocks become fully or overexploited, regulations become more rigid and complex. Fishery managers and fishermen have little or no control over natural production except to change fishing effort and associated fishing mortality.[101] These regulatory efforts can increase fish population abundance and production in some fisheries, but with timing and outcomes that are often uncertain. In the agricultural sector, programs have been developed to help farmers manage financial risks caused by variations in the natural environment such as drought, floods, or disease. To some degree the well-being of producers, communities, lending institutions, and other input sectors are protected by programs such as crop insurance and other types of disaster relief. Some have questioned whether similar programs are needed for the fishing industry, albeit with refinements that consider the unique nature of fisheries.

Many agree that more resources for data collection and stock assessments are needed to improve the understanding of marine ecosystems and marine populations. They reason that through more precise information and by reducing the risk of overfishing, the need for precautionary measures could be lessened and fishing could be increased while remaining within biological limits. However, the benefits associated with marginal improvements in information will decrease with increasing investments in this area. At some point the costs associated with collecting more information become greater than the associated benefits. Regardless of how much data, assessments, and management are improved, constraints to fishing associated with limits to natural production and unpredictable variations in the natural environment will still occur.

Managers must also contend with environmental factors over which they have no control such as climate change and the loss and degradation of fish habitat. Environmental degradation reduces the resiliency of marine and coastal ecosystems and the productivity of marine resources. It also decreases or shifts the geographic range and size of marine populations such as fish and protected species (marine mammals and endangered species). Managers are challenged to find ways to maintain the productivity and health of marine systems and associated populations while minimizing constraints on a variety of economic activities. This may require greater reliance on developing management systems that recognize linkages across government agencies, scientific disciplines, and different oceans activities.

Some would question whether current management institutions can incorporate resource constraints and variability while minimizing disruption of livelihoods and the nature of coastal communities. Some have advocated for catch share programs because of the flexibility they can provide to fishermen. However, many fishermen are skeptical of catch share programs because of issues related to allocation and unanticipated outcomes. Further progress in this area will require integration of social and cultural concerns as well as bioeconomic analysis. Cooperative research and greater industry input during data collection and analysis have also been considered as management costs have increased and information collection requirements have expanded. As marine fisheries evolve many will continue to question the respective roles and investments of public (federal, regional, and state), and private institutions.

APPENDIX A. MSFCMA BACKGROUND

Federal Fisheries Management History

Historically, coastal states managed marine sport and commercial fisheries in nearshore state waters, where most seafood was caught. On September 28, 1945, President Truman issued two proclamations addressing U.S. rights to marine resources beyond the U.S. 3-mile territorial sea.[102] The proclamations expressed the need to conserve and manage living resources and to establish conservation zones in areas of the high seas adjacent to the coasts of the United States. However, the rights proclaimed for fishing did not provide for exclusive jurisdiction over fisheries resources.[103] As fishing technology advanced and market demand increased, fishermen built up fishing capacity, increased catches, and fished more intensively in offshore areas.

In the 1950s and 1960s, increasing numbers of foreign fishing vessels began operating in waters adjacent to the United States. Since the United States only claimed a 3-mile jurisdiction,[104] foreign vessels could fish many of the same stocks caught by U.S. fishermen. U.S. fishermen deplored this "foreign encroachment" and alleged that overfishing was causing stress on, or outright depletion of, fish stocks. United Nations Law of the Sea Treaty negotiations in the early and mid-1970s as well as actions by other coastal nations provided impetus for unilateral U.S. action to declare jurisdiction over fisheries resources within 200 miles of the coastline.[105]

When the United States enacted the Fishery Conservation and Management Act (FCMA, P.L. 94- 265), later renamed the Magnuson Fishery Conservation and Management Act (MFCMA, P.L. 97-191) and more recently the Magnuson-Stevens Fishery Conservation and Management Act (MSFCMA, 109-479), it created a new system of federal fishery management. After several years of debate the FCMA was signed into law on April 13, 1976. On March 1, 1977, marine fishery resources within 200 miles of all U.S. coasts, but outside state waters, came under federal jurisdiction. Initially a substantial portion of the fish caught from federal offshore waters was allocated to foreign fishing fleets. However, the 1980 American Fisheries Promotion Act (Title II of P.L. 96-561) and other FCMA amendments provided incentives to expand domestic fishing and processing industries and decrease foreign catch allocations. On March 10, 1983, the 200- mile fishery conservation zone was superseded by a 200-mile exclusive economic zone (EEZ), proclaimed by President Reagan (Presidential Proclamation 5030).

Management Under the MSFCMA

With the enactment of the FCMA in 1977, an entirely new, regional management system was established to regulate fisheries with priority given to domestic fishermen. Primary federal management authority was vested in the National Marine Fisheries Service (NMFS, also popularly referred to as NOAA Fisheries) within the National Oceanic and Atmospheric Administration (NOAA) of the U.S. Department of Commerce.[106]

The FCMA established eight Regional Fishery Management Councils,[107] which include New England, Mid-Atlantic, South Atlantic, Gulf of Mexico, Caribbean, Pacific (West Coast), Alaska, and Western Pacific. Each council is comprised of marine fishery management agency representatives from each state in the region, the NMFS regional director, and members appointed by the Secretary of Commerce. Appointments are made from lists of candidates knowledgeable about fishery resources that are submitted by state governors.[108] Councils receive input from a variety of advisory committees, species committees, and ad hoc committees. Each council has a Scientific and Statistical Committee (SSC) and, depending on the council, various subcommittees for specific species. The SSC provides the council with scientific advice by developing, collecting, evaluating, and reviewing information during development of fishery management plans and amendments.[109] Members of SSCs include individuals knowledgeable in fisheries from state and federal agencies, universities, and the public.

Fishery management plans (FMPs) are prepared by each council for those fisheries occurring primarily in federal waters and which require active federal management. FMPs consist of management measures and related actions needed to manage stocks such as catch limits, minimum sizes, seasons, closed areas, vessel permitting, and other measures. Public input is a major element of the council process where the public, including fishermen and environmentalists, provides information and comments during the FMP development process.

Most data collection and scientific assessments that support development of FMPs are undertaken by NMFS. Most information is collected and analyzed at NMFS regional science centers and associated laboratories while management functions are conducted from NMFS regional headquarters. After review of the recommendations of appropriate council committees and approval by the council, a proposed action is then submitted to NMFS for review. The review is governed by a strict process that includes additional opportunity for public comment and subsequent approval, partial approval, or disapproval by the Secretary of Commerce. Approved plans are implemented through regulations drafted by NMFS regional management offices and published in the *Federal Register*. These regulations are enforced by NMFS, the Coast Guard, and state fishery enforcement agencies. Plans are amended periodically to account for changes in the fishery and the need for new management approaches.

Together these councils and NMFS have developed and implemented 46 FMPs for various fish and shellfish resources, with additional FMPs and FMP amendments in various stages of development. Some plans are created for an individual species or a few related ones (e.g., FMPs for red drum by the South Atlantic Council and for species of shrimp by the Gulf of Mexico Council).

Others are developed for larger species assemblages inhabiting similar habitats (e.g., FMPs for Gulf of Alaska groundfish by the North Pacific Council and for reef fish by the Gulf of Mexico Council). NMFS manages wide-ranging Atlantic highly migratory species such as tunas, sharks, swordfish and billfish. Many of the implemented plans have been amended (some more than 30 times), and some plans have been developed and implemented jointly by two or more councils.

Fisheries Statistics

The United States has the largest EEZ in the world which includes areas in three oceans (Pacific, Atlantic, and Arctic) with a total area of 3.4 million square nautical miles. These areas also contain some of the most productive fisheries in the world. After passage of the FCMA, foreign catch from the U.S. FCZ declined from about 3.8 billion pounds in 1977 to zero since 1992. Accompanying the decline of foreign catch, domestic offshore catch in federal waters increased dramatically, from about 1.6 billion pounds (1977) to more than 6.3 billion pounds in 1986- 1988.[110] After this peak, annual landings from federal waters have generally ranged from 5 to around 6.5 billion pounds. (*Figure A-1.*)

In 2012, U.S. fishermen landed 10.2 billion pounds of unprocessed fish and shellfish from state, federal, and outside U.S. waters (high seas areas and EEZs of other countries) with a value of $5.63 billion at the dock.[111] Total domestic (state and federal) landings totaled 9.63 billion pounds of which 3.19 billion pounds were from state waters and 6.43 billion pounds were from federal waters. Landings for human food from both state and federal waters totaled 7.48 billion pounds while 2.16 billion pounds were landed for industrial purposes such as animal feeds. The top five species ranked by volume were pollock (2.9 billion pounds), menhaden[112] (1.8 billion pounds), cod (728 million pounds), flatfish (702 million pounds) and salmon (635 million pounds). The value of domestically produced edible products was $9.5 billion while the value of industrial products was $747 million. The top five species by value were crabs ($680 million), scallops ($561 million), shrimp ($490 million), salmon ($489 million), and lobster ($465 million).[113]

In 2012, U.S. per capita consumption of seafood was 14.4 pounds, down 0.8 pounds from 2011.[114] Approximately 91% of the seafood consumed in the United States is imported from other countries.[115] In 2012, imports of edible fishery products were 5.4 billion pounds with a value of $16.7 billion.[116] Exports were 3.3 billion pounds valued at $5.5 billion. Imports are generally composed of relatively high valued species such as shrimp and salmon, and products that have been processed to some degree. U.S. consumers spent an estimated $82.6 billion on edible seafood in 2012 of which $55.2 billion was spent in restaurants and other food service establishments.

In 2012, nearly 9.4 million anglers made approximately 70 million marine recreational fishing trips in the United States.[117] Marine recreational anglers caught an estimated 380 million fish in 2012. Recreational catch that was retained totaled approximately 140 million fish and weighed about 203 million pounds. Most fishing trips were taken on the Atlantic coast (38 million), followed by the Gulf coast (24 million), and the Pacific coast (5.7 million). In 2011, a nationwide survey, conducted every five years, estimated that saltwater recreational anglers spent more than $10.3 billion on their fishing trips and equipment.[118]

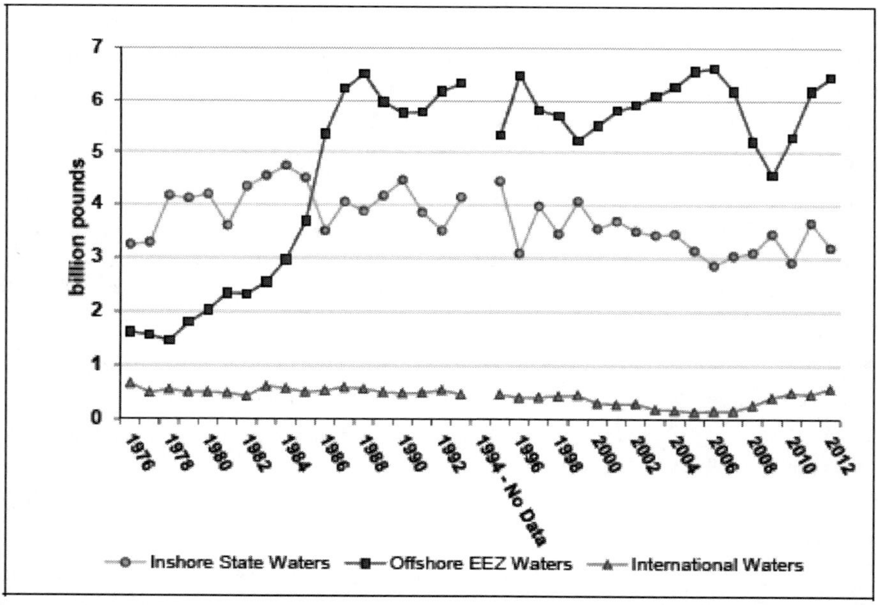

Source: NMFS, Fisheries of the United States (various years), Current Fishery Statistics series.

Note: Total includes both industrial and edible fish and shellfish harvest.

Figure A-1. U.S. Commercial Fish and Shellfish Harvest, 1976-2012.

Constituencies

A variety of groups are involved in managing and utilizing marine fisheries. Although they often share general goals related to maintaining productivity and healthy ecosystems, they often disagree on how to achieve these goals. The following characterizations are general and may vary considerably depending on region and fishery. As Congress considers reauthorization of the MSFCMA, these diverse groups are likely to advocate for a wide range of policies. Advocates of most categories of groups also include national, regional, and local representation as well as individuals or informal groups.

Recreational

Recreational interests include a broad variety of users with different approaches and motivations to fishing. In contrast to commercial fishing, most recreational activities are closely related to the satisfaction that individuals derive from the experience of fishing or observing fish. Generally, activities include a range of often overlapping categories such as non-extractive sports such as snorkeling and scuba diving where utility is gained from observing the marine environment; catch and release fishing where the experience of catching fish is the main objective; and extractive recreational fishing where anglers catch fish for consumption. Different modes of recreational angling include fishing from shore, boats, and charters or party boats. Charters and party boats are commercial operations which provide fishing opportunities to sport anglers for a fee. Recreational interests are often concerned about decisions affecting access to fisheries and allocation of fishing quotas among different interests. Issues and types of recreational interests also vary by region, species targeted, and mode of fishing.

Commercial

Commercial fishing is composed of diverse interests, but it generally involves the harvest of fish for commercial sale. However, in many cases the commercial fishing participants also gain satisfaction from maintaining their livelihood and independence. Commercial fishing can be divided according to target species, scale, and region. For example, even within a specific fishery, commercial users may have different views depending on fishing gear (e.g., trawl, gillnet, pots, or longline), port, or fishing grounds (e.g., inshore or offshore). There are also many fishing dependent businesses that supply fishermen (e.g., ice, nets, vessel maintenance) and businesses that use the harvest such as processors, wholesalers, and others. During the last two decades increasing emphasis has been placed on the inter-dependence between fishing dependent business and maintaining fishing communities. Although national issues often emerge during reauthorizations, regional issues are sometimes reflected in specific provisions of the act. Most fishing associations and representatives are regional in nature and offer regional perspectives.

Environmental

During the last several decades, national and regional environmental groups have gained greater influence in the fishery management process. Initially environmental advocacy focused on indirect harm from the take in fishing gear of marine mammals, sea turtles, and sea birds. Environmental groups expanded their concerns to include overfishing, bycatch, habitat, and marine biodiversity. Interests of environmentalists and fishermen sometimes conflict when fishing may be constrained by greater regulation or coincide for issues such as degradation of habitat from activities other than fishing.

Native American

Many Native American groups are concerned with marine fishing because of their cultural, traditional, and subsistence links to marine resources. In some cases the federal government is required to protect and maintain the treaty rights of some tribes that guarantee access to certain fishery resources. The long-term goals of tribes and indigenous peoples generally include safeguarding cultural traditions, promoting economic stability, and encouraging resource sustainability.

Fishery Scientists and Managers

Fishery scientists and mangers are generally concerned with conserving and managing fishery resources to ensure future benefits. Scientists from academia, the private sector, and state and federal agencies analyze biological, social, cultural, and economic effects of federal fisheries management policy. One of their primary concerns is the availability of adequate funding for data collection, stock assessments, and other analyses necessary to inform fishery managers. Fishery managers develop fishery policies and implement the MSFCMA management measures. Management measures are developed by RFMC members and staff, and NOAA personnel. NOAA personnel also implement management measures by drafting, implementing and enforcing fishery regulations.

Consumers

Consumers are concerned with the availability, quality, and safety of seafood products. There is also growing public concern with making choices that promote sustainable fisheries. A number of non-governmental organizations have developed ecolabels and seafood guides to inform the choices of consumers and seafood buyers. Another growing concern is fraudulent seafood sales and marketing—an act of defrauding buyers of seafood for economic gain, such as mislabeling products or substituting a low valued species for high valued species. Seafood safety and seafood fraud are regulated under the Federal Food, Drug, and Cosmetic Act of 1938 (FFDCA; 21 U.S.C. §§301 et seq.), which is administered by the Food and Drug Administration (FDA).

APPENDIX B. MAGNUSON-STEVENS ACT REAUTHORIZATION IN 109TH CONGRESS

On January 12, 2007, the President signed the Magnuson-Stevens Fishery Conservation and Management Reauthorization Act of 2006 (MSFCMA) (P.L. 109-479, 16 U.S.C. §§1801, et seq.). Major Congressional goals of the 2006 reauthorization included ending overfishing, developing guidelines for catch shares, improving science and the use of science in federal policy decision-making, modifying fishery management council procedures, and enhancing international cooperation. The following summaries focus on amendments that many consider to be most significant.[119]

Conservation and Management

Overfishing and Annual Catch Limits

Section 104 of the MSFCMA mandated the use of annual catch levels (ACLs) to prevent overfishing and to maintain sustainable harvests. Councils are now required to set fishing limits within the range of scientific recommendations. Major changes included the following:

- mandating every fishery management plan establish ACLs at levels such that overfishing does not occur in the fishery;

- requiring ACLs by 2010 for fisheries already subject to overfishing and by 2011 for all other fisheries; and
- requiring councils to develop and implement a rebuilding plan within two years of a stock being declared overfished.

Regional Fishery Management Councils

Section 103 of the legislation increased the role of science in decision-making through a number of provisions focused on the RFMCs' Scientific and Statistical Committees (SSCs) by:

- specifying that the SSCs are to provide their councils with scientific advice needed for management decisions;
- requiring the regional councils to develop five-year research priorities for fisheries, fisheries interactions, habitats, and other areas of research necessary for management;
- modifying regional council fishery management plan procedures, including a requirement to improve coordination of the MSFCMA regulatory process and environmental review under the National Environmental Policy Act (NEPA; 42 U.S.C. §§4321, et seq.) (§107);
- requiring the SSCs to advise the councils on a variety of issues, including stock status and health, acceptable biological catch, overfishing, bycatch, rebuilding targets, habitat status, social and economic impacts, and sustainability of fishing practices; and
- requiring the SSC appointees be federal, state, academic, or independent experts with strong scientific or technical credentials and experience.

Section 103 of the MSFCMA amendments also modified requirements for appointing and training RFMC members. Activities by RFMCs and panels were also modified to enhance transparency of the management process. Provisions of this section sought to inform and strengthen RFMCs by:

- establishing a RFMC training program on the fishery management process for existing council members and new council members; and
- clarifying conflict-of-interest and recusal requirements by ensuring that RFMC members and Scientific and Statistical Committees (SSCs) disclose any financial arrangements that might involve conflict of interest.

Limited Access Privilege Programs

Section 106 of the legislation authorized national guidelines for Limited Access Privilege Programs (LAPPs) commonly known as catch shares, individual transferable quotas (ITQs), or individual fishing quotas (IFQs). LAPPs are federal management systems that divide the total allowable catch of a fishery among fishery participants for their exclusive use. Use of LAPPs is intended to avoid "derby" fishing where many vessels are competing for limited resources. Features of LAPPs, as provided in the act, include the following:

- allows for allocation of harvesting privileges to individuals, corporations, fishing communities, or regional fishery associations;
- allows only fisheries that have been operating under a limited access system[120] to be eligible for management under a LAPP system;
- directs the National Oceanic and Atmospheric Administration (NOAA) to develop criteria for eligibility by considering several factors, such as traditional fishing practices, the cultural and social elements of the fishery, and the severity of projected economic and social impacts of LAPPs;
- allows processors to hold LAPPs and participate in the normal allocation process (but not by allocation of a separate processor quota); and
- requires a formal and detailed review five years after implementation of each program, and thereafter review within every seven years.

Information and Research

Recreational Fisheries

Section 201 required improvements in data collection from recreational fisheries. Programs and priorities included the following:

- establishing a national program to create eight regional registry programs for marine recreational fishermen;
- directing the Secretary to exempt individuals from the regional registry where state programs meet defined criteria; and
- improving the Marine Recreational Fishery Statistics Survey.

Cooperative Research

Section 204 of the MSFCMA amendments directed the Secretary to establish a cooperative research and management program. Requirements of this program include

- identifying projects to be competitively funded by the Secretary, with priority given to collecting data to improve stock assessments, assessing bycatch and mortality, designing technology to reduce bycatch, identifying important habitat, and collecting social and economic data;
- directing the Secretary of Commerce to establish guidelines to ensure that participation does not result in the loss of a participant's catch history or unexpended days-at-sea as part of a limited entry system; and
- requiring the Secretary of Commerce to issue regulations for expediting regionally based experimental fishery permits.

Regional Ecosystem Research

Section 210 of the 2006 amendments focused on developing information to implement pilot projects by:

- requiring the Secretary of Commerce, in consultation with the councils, to study the state of the science for advancing the concepts and integration of ecosystem considerations in regional fishery management;
- including recommendations in the study for scientific data, information, and technology requirements for understanding ecosystem processes, and methods for integrating information from different sources;
- incorporating broad stakeholder participation, recommendations to account for environmental variation, and a description of existing and new efforts to implement ecosystem approaches; and
- providing technical advice and assistance to councils for developing and designing regional pilot programs.

Other Research

Other research efforts include

- establishing a deep sea coral research and technology program (§211); and
- researching and promoting new gear technologies to further reduce bycatch (§116);

International Conservation and Management

Illegal, Unreported, and Unregulated Fishing

Sections 401-403 of the MSFCMA amendments included provisions to strengthen the ability of international fishery management organizations and the United States to appropriately enforce conservation and management measures for high seas fisheries. Provisions included

- undertaking activities to improve international compliance and monitoring of high seas fisheries, and report to Congress on progress in reducing illegal, unreported, and unregulated fishing (IUU) (§401);
- strengthening the ability of fishery management organizations to stop IUU fishing (§403);
- requiring the Secretary of Commerce to define IUU, and specifying that the definition must include violations of quotas or other rules established by international agreement (§403); and
- allowing for the use of measures authorized under the High Seas Driftnet Act to force compliance in cases where regional or international fishery management organizations are unable to stop IUU fishing (§403).

Western and Central Pacific Fisheries Convention (WCPFC)

Title V (§§501-511) implements provisions of the WCPFC by:

- providing for U.S. participation in the Western and Central Pacific Fisheries Commission such as appointment of commissioners, and other administrative matters;
- defining the authorities of the Secretary of State and Secretary of Commerce; and
- making conservation and management measures adopted by the WCPF Commission legally binding upon nations and vessels subject to U.S. jurisdiction.

Other Provisions

Additional provisions included the following:

- authorizing appropriations of $338.8 million in FY2007 with a $9.84 million annual increase for implementing activities through FY2013 (§7);
- establishing marine education and training programs for Western and North Pacific communities to improve communication, education, and training on marine resource issues and increase scientific education opportunities for marine-related professions among coastal community residents (§109);
- establishing regional economic transition programs to provide disaster relief at the request of Governors of affected states (§113); and
- implementing the Pacific Whiting Act of 2006 between the United States and Canada (§§601-611).

NMFS has summarized various activities associated with implementing *P.L. 109-479* in a table that lists tasks.[121] Examples of implementation activities include (1) a report by NMFS to Congress on implementing new provisions relating to reduce illegal, unreported, and unregulated (IUU) fishing activities[122] and (2) final guidance amending National Standard 1, designed to end overfishing through new requirements for annual catch limits and other accountability measures.[123] In addition, NMFS released a new national policy on the use of catch shares in fishery management plans.[124]

APPENDIX C. ACRONYMS

ACLs	Annual Catch Limits
AFF	Asset Forfeiture Fund
AMs	Accountability Measures
EEZ	Exclusive Economic Zone
ESA	Endangered Species Act
FMP	Fishery Management Plan
IFQs	Individual Fishing Quotas
ITQs	Individual Transferable Quotas
IUU	Illegal, Unreported, and Unregulated

LAPPs	Limited Access Privilege Programs
MSY	Maximum Sustainable Yield
NEPA	National Environmental Policy Act
NRC	National Research Council
OY	Optimum Yield
RFMC	Fishery Management Council
RFMO	Regional Fishery Management Organization
SSC	Scientific and Statistical Committee

End Notes

[1] A fishery is defined in the MSFCMA as (1) one or more stocks of fish which can be treated as a unit for the purposes of conservation and management and which are identified on the basis of geographical, scientific, technical, recreational, and economic characteristics; and (2) any fishing for such stocks.

[2] House Committee on Natural Resources, Discussion Draft, December 2013, http:// natural resources.house.gov/uploadedfiles/magnusonstevensactdiscussiondraft-113. pdf.

[3] Molly Dischner, "Senate draft of fisheries act begins circulating," Alaska Journal of Commerce, April 17, 2014.

[4] The MSFMCA defines the term "stock of fish" as a species, subspecies, geographical grouping, or other category of fish capable of management as a unit.

[5] There are exceptions for ecosystem component stocks.

[6] MSFCMA requires the Secretary of Commerce to establish advisory guidelines for councils based on the national standards. The statute explicitly states that guidelines "shall not have the force and effect of law." Guidelines are meant to provide guidance to councils as they develop FMPs or FMP amendments and they do not mandate specific management measures for any fishery.

[7] NOAA Fisheries, Summary of Comments received on the Advance Notice of Proposed Rulemaking on potential adjustments to the National Standard Guidelines, Silver Spring, MD.

[8] U.S. Dept. of Commerce, National Oceanic and Atmospheric Administration, "Magnuson-Stevens Act Provisions; Annual Catch Limits; National Standard Guidelines," 74 Federal Register 3178-3213, January 16, 2009. Hereafter cited as Final NOAA Guidelines 2009.

[9] NOAA Fisheries, Status of Stocks 2013, Annual Report to Congress on the Status of U.S. Fisheries, Silver Spring, MD, http://www.nmfs.noaa.gov/sfa/fisheries_eco/status_of_fisheries/archive/2013/status_of_stocks_2013_web.pdf.

[10] Ocean Studies Board, Evaluating the Effectiveness of Fish Stock Rebuilding Plans in the United States, National Research Council, Washington, DC, 2013.

[11] Carl Safina, Andrew Rosenberg, and Ransom A. Myers, et al., "U.S. Ocean Fish Recovery: Staying the Course," Science, vol. 309 (July 29, 2005), pp. 707-708.

[12] For a discussion of stock rebuilding timeframes and current exceptions see pages 3200-3201 of U.S. Dept. of Commerce, National Oceanic and Atmospheric Administration, "Magnuson-Stevens Act Provisions; Annual Catch Limits; National Standard Guidelines," 74 Federal Register 3178-3213, January 16, 2009.

[13] Senator Kristen Gillibrand, "Schumer, Gillibrand Reintroduce Legislation to Help Save Long Island's Fishing Industry by Increasing Flexibility in Arbitrary Federal Fishing Rules," press release, March 23, 2011, http://www.gillibrand.senate.gov/ newsroom/ press/ release/schumer-gillibrand-reintroduce-legislation-to-help-savelong-islands-fishing-industry-by-increasing-flexibility-in-arbitrary-federal-fishing-rules.

[14] Sherry L. Larkin, Gil Sylvia, and Michael Harte, et al., "Optimal Rebuilding of Fish Stocks in Different Nations: Bioeconomic Lessons for Regulators," Marine Resource Economics, vol. 21 (2007), pp. 395-413.

[15] U.S. Congress, House Committee on Natural Resources, Hilborn Presentation on Magnuson-Stevens Reauthorization, Magnuson-Stevens Fisheries Conservation and Management Act, 113th Cong., 1st sess., September 11, 2013. Hereafter cited as Hilborn 2013.

[16] Hilborn 2013.

[17] Fished at rates below MSY.

[18] Maximum Economic Yield is achieved when the difference between the costs of fishing and revenue is greatest (net benefits are maximized). Generally, this occurs because the marginal benefits of the fishing fleet begin decreasing before MSY is reached while the marginal costs of fishing remain constant.

[19] Organization for Economic Co-operation and Development, "Managing the transition: Distributional issues of fish stock rebuilding," in The Economics of Rebuilding Fisheries (OECD , 2010), pp. 141-166.

[20] When property rights are more secure it often provides users with incentives to give greater weight to future outcomes.

[21] It may be difficult to define "traditional" because fishing fleets are often composed of a continuum of vessel sizes technology, and fishing strategies.

[22] Recent adoption of catch shares in some fisheries has also modified the nature of fishing and fishing communities.

[23] Technical innovations such as gear modifications and area management have improved the ability of managers and fishermen to target specific species.

[24] U.S. Congress, Senate Committee on Commerce, Science, and Transportation, Subcommittee on Oceans, Atmosphere, Fisheries, and Coast Guard, Testimony of Trevor A. Branch, Ph. D. Assistant Professor, University of Washington, West Coast and Western Pacific Perspectives on Magnuson-Stevens Act Reauthorization, 113th Cong., 2ad sess., January 30, 2014.

[25] Final NOAA Guidelines 2009.

[26] The mixed stock exception may be used if (1) the action will provide net benefits to the nation; (2) the action provides results that cannot be achieved by other means; and (3) the rate of fishing mortality will not cause the stock or stock complex (a grouping of often similar species) to fall below minimum stock size threshold more than 50% of the time.

[27] Management and scientific uncertainty were used in the guidelines to describe the sources of uncertainty in fisheries. However, typically the terms systematic and random are used to describe two categories of uncertainty. Systematic uncertainties consistently cause data values to be too small or too large while random uncertainties occur without a predictable pattern. Accuracy describes very small systematic errors (refers to how closely the measured value of a quantity corresponds to its true value). Precision describes small random errors (refers to agreement among repeated measurements). Fishery science is subject to both types of uncertainty, but systematic errors are often more difficult to account for such as potential under-reporting of bycatch.

[28] Often measures such as net mesh size, closed areas, and minimum sizes were used instead of quotas to regulate fisheries instead of quotas.

[29] Alec D. MacCall, Dynamic Geography of Marine Fish Populations (WA: University of Washington, 1990), pp. 2-3.

[30] NOAA Fisheries, Status of the Stocks 2013, Annual Report to Congress on the Status of the U.S. Fisheries, Silver Spring, MD, 2013. http://www.nmfs.noaa.gov/sfa/fisheries_eco/status_of_fisheries/archive/2013/ status_of_stocks_2013_web.pdf.

[31] U.S. Congress, House Committee on Natural Resources, Testimony of Robert G. Hayes, Legislative Hearing on H.R. 594, H.R. 1013, H.R. 1646, H.R. 2304, H.R. 2610, H.R. 2753, H.R. 2772, and H.R. 3061, 112th Cong., 1st sess., December 1, 2011.

[32] Northeast Fisheries Science Center, 53rd Northeast Regional Stock Assessment Workshop (53rd SAW) Assessment Report., National Marine Fisheries Service, U.S. Dept. of Commerce, Northeast Fisheries Science Center Reference Document 12-05, Woods Hole, MA, March 5, 2012, http://www.nefsc.noaa.gov/nefsc/publications/.

[33] AMs are used to compensate when landings are greater than the quota for the fishing year. For example, if 10 tons more than the quote is harvested, 10 tons will be subtracted from the quota for the next fishing year.

[34] Currently "ecosystem stocks" as defined in the NOAA Guidelines, is more restrictive than the definition used in these bills.

[35] Each regional fishery management council has an SSC. The SSC provides the council with scientific advice by developing, collecting, evaluating, and reviewing information during the development of fishery management plans and amendments.

[36] 16 U.S.C. §1851(26).

[37] U.S. General Accounting Office, Individual Fishing Quotas, GAO-03-159, December 2002, http://www.nmfs.noaa.gov/sfa/domes_fish/catchshare/docs/GAO-03-159.pdf.

[38] Ayeisha A. Brinson and Eric M. Thunberg, The Economic Performance of U.S. Catch Share Programs, National Oceanic and Atmospheric Administration, NOAA Technical Memorandum NMFS-F/SPO-133, August 2013.

[39] Courtney Carothers, Daniel K. Lew, and Jennifer Sepez, "Fishing rights and small communities: Alaska halibut IFQ transfer patterns," Ocean and Coastal Management, vol. 53, no. 9 (September 2010), pp. 518-523.

[40] Before access was limited, many fisheries were overcapitalized with greater investments in vessels and gear than needed to harvest available resources.

[41] In the 112th Congress bills included H.R. 1646, H.R. 2772 while in the 113th Congress S. 221 was introduced.

[42] H.R. 2772 only would have terminated new programs while S. 1678 and S. 221 would have terminated both existing and new programs.

[43] U.S. Congress, House Committee on Natural Resources, Testimony of Robert G. Hayes, Legislative Hearing on H.R. 594, H.R. 1013, H.R. 1646, H.R. 2304, H.R. 2610, H.R. 2753, H.R. 2772, and H.R. 3061, 112th Cong., 1st sess., December 1, 2011.

[44] To prohibit or require specific conditions for management measures perceived as extreme such as closing fisheries.

[45] The bill would also set a deadline for secretarial decisions on disaster declarations and modify criteria for limited access privilege program approval.

[46] Ecosystem Principles Advisory Panel, Ecosystem-Based Management: A Report to Congress, National Marine Fisheries Service, November 1998.

[47] Department of Commerce, National Oceanic and Atmospheric Administration, and National Marine Fisheries Service, The state of Science to Support an Ecosystem Approach to

Regional Fishery Management, Pursuant to the Magnuson-Stevens Fishery conservation and Management Act, Section 406(f), April 2009, http://www.nmfs.noaa.gov/msa2007/docs/tm_96_repto_congress_final.pdf (hereinafter cited as State of Science 2009).

[48] State of Science 2009, p.

[49] Pacific Fishery Management Council, Managing Our Nation's Fisheries, Proceedings of a conference on fisheries management, Portland, OR, 2013, p. 152, http://www.managing fisheries.org/2013_documents/ Proceedings_complete.pdf. Hereafter cited as Managing Our Nation's Fisheries 2013.

[50] Managing Our Nation's Fisheries 2013, p. 154.

[51] Managing Our Nation's Fisheries 2013, p. 154.

[52] 16 U.S.C. §1802(10).

[53] Managing Our Nation's Fisheries 2013, p. 253.

[54] Managing Our Nation's Fisheries 2013.

[55] 16 U.S.C. §1855(b)

[56] 16 U.S.C. §1851(a).

[57] 16 U.S.C. §1802(2).

[58] 16 U.S.C. §1802(9).

[59] 16 U.S.C. §1802(38).

[60] 16 U.S.C. §1862 and 16 U.S.C. §1865

[61] National Marine Fisheries Service, U.S. National Bycatch Report, NOAA Tech. Memo. NMFS-F.SPO-117C, 2011, http://www.nmfs.noaa.gov/by_catch/BREP2011/2011_National_Bycatch_Report.pdf.

[62] Lee R. Benaka, Laura F. Cimo, and Lekeliad D. Jenkins, "Bycatch Provisions in the Reauthorized Magnuson-Stevens Act," Marine Fisheries Review, vol. 74, no. 2 (2012), pp. 1-10.

[63] Pacific Fishery Management Council, Reducing Bycatch in Alaska, http://www.npfmc.org/wp-content/ PDFdocuments/bycatch/Bycatchflyer913.pdf.

[64] Amadnad Keledjian et al., Wasted Catch: Unsolved Problems in U.S. Fisheries, Oceana, March 2014, http://oceana.org/sites/default/files/reports/Bycatch_Report_FINAL.pdf.

[65] In some cases such as the Long Island New York hard shell clam fishery, Northern Mariana Islands fisheries— following a super typhoon, and the Florida shark fishery—it took two to three years for the Secretary to make a determination.

[66] 16 U.S.C. 1881a(b)

[67] National Oceanic and Atmospheric Administration, "Confidentiality of Information; Magnuson-Stevens Fishery Conservation and Management Reauthorization Act," 77 Federal Register 100, May 23, 2012.

[68] 36 C.F.R. §242.4

[69] Pacific Fishery Management Council, Managing Our Nation's Fisheries, Proceedings of a conference on fisheries management, Portland, OR, 2013, p. 272.

[70] Courtney Carothers, Daniel K. Lew, and Jennifer Sepez, "Fishing rights and small communities: Alaska halibut IFQ transfer patterns," Ocean and Coastal Management, vol. 53, no. 9 (September 2010), pp. 518-523.

[71] The determination involves a positive or negative certification. A positive certification is issued if the nation has addressed the activities for which it was certified while a negative certification may result in denial of port access and potential import restrictions.

[72] A nearly identical bill (S. 1980) was introduced in the 112th Congress and reported by the Senate Committee on Commerce, science and Transportation, but no further action was taken.

[73] Originally the MSFCMA was named the Fishery Conservation and Management Act.

[74] H.R. 1430 and S. 681 include both the Southeastern states and the Gulf while S. 747 only includes the Gulf of Mexico.

[75] The Gulf States Marine Fisheries Commission is a regional management body that coordinates management of state marine and anadromous species.

[76] Section 311(e)

[77] U.S. Department of Commerce, Office of Inspector General, Review of NOAA Fisheries Enforcement Asset Forfeiture Fund, Office of Investigations, Final Report No. OIG-19887-1, July 2010, http://www.oig.doc.gov/ OIGPublications/2010.07.01-IG-to-NOAA.pdf.

[78] National Oceanic and Atmospheric Administration, "Timeline of NOAA Enforcement Program Improvements," http://www.noaa.gov/lawenforcementupdates/timeline.html.

[79] National Marine Fisheries Service, Gulf of Mexico Fishery Management Plan Frequently Asked Question, January 2013.

[80] Jane Chalmers, Authority for Aquaculture under the Magnuson-Stevens Act, National Oceanic and Atmospheric Administration, Washington, DC, June 9, 2011.

[81] Cathy Roheim Wessells, Kevern Cochrane, and Carolyn Deere, et al., Product certification and eco-labeling for fisheries sustainability, Food and Agriculture Organization of the United Nations, FOA Fisheries Technical Paper 422, Rome, Italy, 2001, http://www.fao.org/docrep/005/y2789e/y2789e00.htm. Hereafter cited as FAO 2001.

[82] Nicolai V. Kuminoff, Darrell J. Bosch, and Dan Kauffman, et al., "The Growing Supply of Ecolabeled Seafood: An Economic Perspective," Sustainable Development and International Law & Policy, vol. 9, no. 25 (Fall 2008). Hereafter cited as Kuminoff 2008.

[83] Guides are often also categorized by color codes such as from red, (avoid) to green, (best choice).

[84] Tracy M. Roberts, "The Rise of Four Institutions: Voluntary Standards, Certification, and Labeling Schemes," Ecology Law Quarterly, vol. 40, no. 107 (2013).

[85] Food & Water Watch, De-Coding Seafood Eco-Labels: Why We Need Public Standards, November 2010, http://www.foodandwaterwatch.org/reports/de-coding-seafood-eco-labels-why-we-need-public-standards/. Hereafter cited as FWW 2010.

[86] The national standards (Section 301(a)) cover a variety of fishery management objectives, including consideration of fishing communities and safety at sea.

[87] Vangelis Vitalis, Private Voluntary Eco-labels: Trade Distorting, Discriminatory and Environmentally Disappointing, Organization for Economic Development and Co-operation, background paper produced for a round table on sustainable development, Paris, France, 2002, http://www.oecd.org/sd-roundtable/papersandpublications/ 39362947.pdf. Concern with potential barriers to trade is another potential problem with certification programs cited by Vitalis.

[88] National Institute for Occupational Safety and Health, Commercial Fishing Safety, Center for Disease Control, 2012, http://www.cdc.gov/niosh/topics/fishing/.

[89] Eric Christensen and Jack Kemerer, "Fishing Vessel Safety," The Coast Guard Proceedings of the Marine Safety and Security Council, Winter 2011.

[90] Letter from Safe at Sea Network to Samuel D. Rauch III, Acting Assistant Administrator for Fisheries, August 2, 2013, http://www.edf.org/sites/default/files/SafeAtSea. LettertoNMFS. FINAL2_.pdf.

[91] NOAA, "Revision of National Standard 10 Guidelines," 76 Federal Register 22343, April 21, 2011.

[92] The NOAA website for ongoing revisions to National Standard 10 can be accessed at, http://www.nmfs.noaa.gov/sfa/ laws_policies/national_standards/ns10_revisions.html.

[93] Department of Commerce, National Oceanic and Atmospheric Administration, National Marine Fisheries Service, National Environmental Policy Act Compliance for Fishery Management Actions under the Magnuson-Stevens Act, February 19, 2013.

[94] Letter from Regional Fishery Management Councils Coordination Committee to Sam Rauch, NMFS Acting Assistant Administrator, March 11, 2013.

[95] For a detailed review of the legislative history of the NMSA, see William L. Chandler and Hannah Gillelan, "The History and Evolution of the National Marine Sanctuaries Act," Environmental Law Reporter, no. 34 (2004): 10506- 10565.

[96] National Oceanic and Atmospheric Administration, NOAA's Regulation of Fishing in National Marine Sanctuaries, July 2008, http://sanctuaries.noaa.gov/library/ pdfs/noaa_regs_nmsfishing_2008.pdf.

[97] 16 U.S.C. §§431-433.

[98] For example, fishing may be regulated or prohibited in national monuments as in the case of the Papahānaumokuākea Marine National Monument (Northwestern Hawaiian Islands Marine Monument where fishing was phased out during the first five years after the monument's establishment.

[99] See CRS Report RS20902, National Monument Issues, by Carol Hardy Vincent.

[100] Jeff Brax, "Zoning the Ocean: Using the National Marine Sanctuaries Act and the Antiquities Act to Establish Marine Protection Areas and Marine Reserves in America," Ecology Law Quarterly, v. 29 (2002): 71-129.

[101] Habitat may be enhanced or restored, but often the causes of habitat loss or degradation are outside the authority of fisheries laws.

[102] "Proclamation No. 2667, Policy of the United States with Respect to the Natural Resources of the Subsoil and Sea-Bed of the Continental Shelf," 10 Federal Register 12303, September 28, 1945 and "Proclamation No. 2668, Policy of the United States with Respect to Coastal Fisheries in Certain Areas of the High Seas," 28 Federal Register 12304, September 28, 1945.

[103] Robert W. Smith, Exclusive Economic Zone Claims: An Analysis and Primary Documents (Dordrecht, The Netherlands: Martinus Nijhoff Publishers, 1986), p. 25.

[104] Subsequently in 1964, P.L. 88-308 prohibited fishing by foreign-flag vessels within 3 miles of the coast; in 1966, P.L. 89-658 proclaimed an expanded 12-mile exclusive U.S. fishery jurisdiction.

[105] The United Nations Convention on the Law of the Sea has not been ratified by the U.S. Senate.

[106] NMFS programs are described in detail at http://www.nmfs.noaa.gov/.

[107] Links to individual council websites are available at http://www.nmfs.noaa.gov/councils/.

[108] For the 2010 report to Congress on council membership, see http://www. nmfs.noaa. gov/sfa/reg_svcs/ Council_Reportocongress/2010ApportionmentReportToCongress.pdf.

[109] Most basic data are collected by NMFS or through agreements with state management agencies and most stock assessments are conducted by NMFS fisheries science centers.

[110] This total includes both landings for human food and landings for industrial purposes (e.g., bait and animal food, reduction to meal and oil, etc.).

[111] U.S. Department of Commerce, National Oceanic and Atmospheric Administration, National Marine Fisheries Service, Fisheries of the United States 2012, Silver Spring, MD, September 2013. Hereafter cited as Fisheries of the United States 2012.

[112] An industrial fishery with products which include fish oils and fish meal.

[113] Most of these products such as crabs are composed of species groups as reported by NMFS.

[114] Fisheries of the United States 2012.

[115] National Marine Fisheries Service, FISHWATCH U.S. Seafood Facts, http://www.fishwatch.gov/farmed_seafood/ outside_the_us.htm.

[116] Fisheries of the United States 2012.

[117] Fisheries of the United States 2012.

[118] U.S. Department of the Interior, U.S. Fish and Wildlife Service, U.S. Department of Commerce, and U.S. Census Bureau, 2011 National Survey of Fishing, Hunting, and Wildlife-Associated Recreation, 2014, http://www.census.gov/ prod/2012pubs/fhw11-nat.pdf.

[119] For more information concerning 2006 amendments to the act and implementation see the website, Magnuson-Stevens Fishery Conservation and Management Act Requtho4rized, Implementing the Magnuson-Stevens Act at http://www.nmfs.noaa.gov/ msa2007/ implementation.htm.

[120] Limited access systems limit fishing participation to existing permits and does not allow for new permits to be issued in the fishery.

[121] NOAA Fisheries, Magnuson-Stevens Fishery Conservation and Management Act Reauthorized, List of implementation priorities, http://www.nmfs.noaa.gov/ msa2007/index.html.

[122] U.S. Department of Commerce, Report to Congress Pursuant to Section 403(a) of the Magnuson-Stevens Fishery Conservation and Management Reauthorization Act of 2006, Washington DC, January 2013, http://www.nmfs.noaa.gov/ia/iuu/msra_page/ 2013_ biennial_report_to_congress__jan_11__2013__final.pdf.

[123] National Oceanic and Atmospheric Administration, "Magnuson-Stevens Act Provisions; Annual Catch Limits; National Standard Guidelines; final Rule," 74 Federal Register 3178-3213, January 16, 2009.

[124] U.S. Department of Commerce, National Oceanic and Atmospheric Administration, NOAA Catch Share Policy, http://www.nmfs.noaa.gov/sfa/management/catch_shares/about/ documents/noaa_cs_policy.pdf.

INDEX

C

D

E

F

G

H

I

J

K

L

M

N

O

S

T

U

V

W

Y